CAD in reinforced concrete detailing and structural steelwork

CAD in reinforced concrete detailing and structural steelwork

Proceedings of a conference organized by the Institution of Civil Engineers and held in London on 26 November 1987

Thomas Telford, London

Conference organized by the Institution of Civil Engineers

Organizing Committee: Dr G. Storer (Chairman), Mr I. Hamilton, Mr J. J. Pike and Dr S. Port

First published 1988

British Library Cataloguing in Publication Data
CAD in reinforced concrete detailing and structural steelwork.
1. Civil engineering structures. Design. Applications of computer system
I. Institution of Civil Engineers
624.1'771'0285

ISBN 0 7277 1340 X

Published for the Institution of Civil Engineers by Thomas Telford Ltd, 1 Heron Quay, London E14 9XF.

Printed in England by Hartnolls Ltd, Bodmin, Cornwall.

Contents

Paper 1. The BARD system for reinforcement detailing — the next 50 000 drawings

M.R. HOLLINGTON, BSc(Eng), PhD, DICE, MICE, MIStructE,
W.G. BATCHELOR, BTech(Eng), and P.W. NOBLE, BSc, MBCS,
Cement & Concrete Association

SYNOPSIS. The BARD system developed by the Cement & Concrete Association is used to produce high quality bar schedules and reinforcement drawings on a wide range of computer equipment. It was designed to increase the productivity of draughtsmen in the detailing of reinforced concrete and is marketed by Wexham Developments Limited.

SYSTEM SPECIFICATION

1. The Cement & Concrete Association started development work on BARD in the mid-1970's with co-operation and feedback from three structural consulting organisations. It was agreed that the concept should be 'single element detailing' whereby the output was to consist of say one beam span or one slab panel presented on an A3 sheet, complete with a bar-schedule. It was also stipulated that the system should be capable of running on low-cost equipment in addition to the more costly mini-computers. At the time it was clear that advances in computer technology were taking place very rapidly and it was essential to design a system capable of moving with these advances. This was important for the following reasons.

(a) First of all, from the users point of view, there is nothing more frustrating than discovering that the software you are using is obsolete merely because the computer on which it runs is obsolete. Or, just as bad, that the software would function more efficiently on a more up to date model if only it could be transferred. The software therefore had to be 'portable'. A hidden cost often overlooked when purchasing software is the cost of training people to use it. The software writer therefore has a responsibility to the user to ensure system portability and so avoid dependence on a particular make of computer.

(b) Similarly, on the output side, there was a requirement to be able to interface the system with a range of plotting devices. In the past most users had opted for drum or flat-bed plotters using ink pens on tracing paper in order to emulate manual drawing office standards. However, the cost of fast drum plotters was often in excess of ten thousand pounds which only some of the larger firms could afford. Consequently, it was important to be able to communicate with less costly devices such as dot matrix printers on the basis that there were many small firms of consulting engineers who could not afford expensive equipment. An interesting observation is that, whereas a year or two ago very few firms were willing to accept dot matrix output, more and more are coming to realise that these printers are a cheap, reliable and fast way of producing drawings.

(c) Finally, from the software supplier's point of view, the cost of writing, documenting and maintaining software is always prohibitive so it is more lucrative if it runs on a wide range of computers.

2. BARD was therefore written in standard ANSI Fortran, communicating with various operating systems and plotting devices through a series of link routines. Consequently, BARD is now running of Prime, Vax, HP3000, Data General and a wide range of micro computers using the MSDOS operating system (IBM PC XT/AT, Apricot, Amstrad PC1512, Olivetti M24/M28.. etc). The cheapest hardware configuration on which BARD will run satisfactorily is currently an Amstrad PC1512 outputting to an Epson EX1000 dot-matrix printer - costing around fifteen hundred pounds. This should be compared with the so-called low cost system used for BARD in 1980 which consisted of a DEC LSI 11/03 microcomputer driving a Benson 1202 plotter which totalled seventeen thousand pounds!

3. To summarise, BARD has been carefully designed to give the detailer a flexible tool that can deal quickly and efficiently with so-called 'bread and butter' applications. BARD cannot produce large general arrangement drawings or some of the more difficult 'one-offs' since it was felt that these tasks did not make the most economical use of the computer. Nevertheless, should the need arise, BARD can be linked with CAD systems by means of an intermediate plotfile although the economics of doing so are debatable.

DATA INPUT

4. Input to BARD is usually via a conversational 'questions and answers' procedure whereby the system builds a datafile, checking each item against the relevant detailing rule. Most questions include a default answer which the user may optionally accept by a single keystroke. These default values are held in 'Standards Files' which can be modified to suit the requirements of a particular Code of Practice. Standards files supplied with the system at present satisfy BS8110. It is not unusual for users to take as little as four or five minutes to input the data required for a typical beam span or slab panel, another five minutes being required to produce the drawing. In practice detailers can produce 20-30 BARD details

per day giving an overall efficiency in excess of 10:1 when compared with manual methods. The resulting output is presented on an A3 sheet of paper and includes both a fixing detail and a weighted bar schedule. A typical BARD drawing, reduced from A3 size, is shown below in Figure 1.

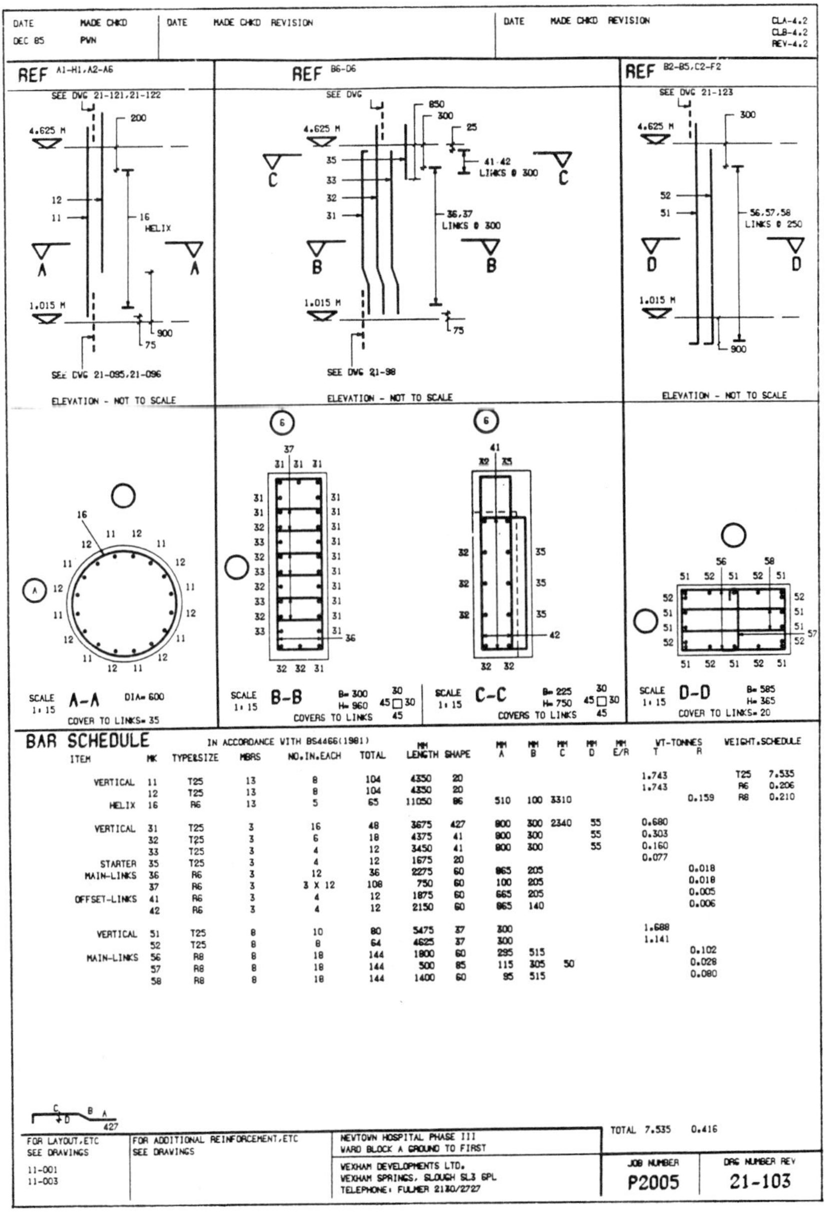

Figure 1

The training period for BARD is only two or three days since the data input is aimed specifically at detailers, no knowledge of CAD being required.

5. It is very difficult to fully automate r.c. detailing by feeding design output directly into the detailing system since most jobs require the expertise of a detailer. However, a link between BARD and the Ove Arup OASYS design suite is available for some applications. In these circumstances the linking software generates BARD data files automatically.

DRAWING OUTPUT

6. In order to make allowance for the very wide range of plotting devices available today, BARD generates ASCII plot files from the conversationally produced data files. The user is then presented with a plot menu from which a device is selected. Using this procedure, electrostatic printers, flat-bed and drum plotters, dot-matrix printers, laser printers and a variety of screen graphics devices have been used to produce BARD drawings. The plotfile is also convenient as a means of communicating with other drafting systems. Some firms insist on all plotted output being channelled through a standard interface which may be linked to their in-house drafting facility. The economics of insisting that 100% of all drawings must be produced by the computer is very questionable however and it often makes more sense to add the odd 5% manually. In these circumstances, BARD offers a facility whereby information relating to bars added manually may be included in the weighted bar-schedule. Alternatively, to deal with elements for which no BARD program is currently available BARDSHED may be used.

BARD PROGRAMS

7. The BARD system consists of the following suite of ten programs:-

BEAMS1 This program details a single beam span, the span bars and support bars being presented on separate elevations. The arrangement of bars follow the recommendations set out in the Concrete Society's "Standard Reinforced Concrete Details", which state that neither the bottom span bars nor the top span bars extend to the column. Continuity through the column is provided by top and bottom support bars which form a splice with the span bars inside the beam cage. Thus the span cage can be completely prefabricated and lifted into position between the edges of support.

BEAMS2 This program details a single beam span where the support bars can form an integral part of the total beam cage, allowing bottom bars to lap at supports if required. Laps and clearances can be achieved by specifying joggled bars.

COLUMNS This program covers columns of square, rectangular

and circular cross-section and caters for cases where the size of the column above that being detailed has the same or reduced section. Up to four columns may be detailed on a single A3 sheet.

FLAT SLABS This program covers the detailing of solid reinforced concrete rectangular flat slab panels which span between column supports.

SOLID SLABS This program covers the detailing of solid rectangular concrete slab panels which span between line supports such as beams or walls. Three types of slab are dealt with as follows:-

(a) Two-way spanning continuous panels
(b) One-way spanning continuous panels
(c) Single one-way spanning simply supported panels

WAFFLE SLABS This program covers the detailing of reinforced concrete rectangular waffle slab panels which span between column or line supports.

TROUGH SLABS This program covers the detailing of reinforced concrete rectangular trough slab panels which have ribs spanning in one direction between line supports.

LOADBEARING WALLS This program covers the detailing of wall panels with openings and solid walls up to 200 metres in length. Nibs, slab starter bars, spacer bars and links can all be provided.

PAD BASES & PILE CAPS This program covers the detailing of rectangular pad bases and rectangular or A-Form pile caps. Column starters can optionally be provided.

BARDSHED This program is used to schedule elements for which no BARD program is currently available. It enables the user to produce bar-schedules more efficiently than by manual methods and also offers the option of presenting the output in A3 format for compatibility with other BARD programs.

BARD IN ACTION

8. BARD is currently being used in nearly fifty offices around the country and, to date, more than 50,000 drawings have been produced.

9. The scale of jobs on which the system has been used vary from large hospitals to small housing projects. One very popular application, not considered at the outset, is the detailing of ground beams for housing schemes where made-up ground has to be utilised. The data files required by BARD are very simple in these circumstances and can often be generated automatically direct from the design phase. Grove Structural

Consultants have used this procedure very successfully for a number of Wimpey Homes projects.

10. Austin Trueman Associates, who have fifteen engineers in their design team, purchased Beams1, Beams2 and Bardshed in April of this year and immediately recovered the costs on a job consisting of only 74 BARD drawings.

11. Alan Marshall Partnership have used BARD for several years and have produced in excess of 5000 drawings. Very few queries are received from site and steel fixers are reported to like the system since it simplifies their task considerably. Some of the Alan Marshall projects on which BARD has been used are listed below:-

Bromsgrove & Redditch District General Hospital
Stoke District General Hospital (Surgical accommodation)
Wordsley Maternity Hospital
Rugby St. Cross Pathology Block
Rugby St. Cross - Phase 3
Leicester Royal Infirmary - Phase 4
Litchfield & Tamworth District General Hospital
King Khalid Mosque - Riyadh

12. Module 2, based at Bridgend, have recently used BARD successfully on a large car-park at Brighton. Module 2 are newcomers to BARD but apparently had no difficulty in applying it to this job, producing 540 drawings via one detailer in a very short time indeed. They intend using BARD on the adjacent hotel complex in the near future.

THE FUTURE

13. When you write software which is to be made available commercially, there is often a conflict of interests between using the very latest technology and the price at which you can market the product. There is also a subtle difference between setting out to produce a saleable package from the outset, and developing a system for in-house use which you subsequently decide might be a saleable product.

14. In the former case, particularly with r.c. detailing in mind, the majority of firms of Consulting Engineers in this country consist of small practices with about ten personnel where the Partners are usually reluctant to spend vast sums on computer programs and equipment.

15. Nevertheless, at the system design stage of BARD, we were very conscious of the developments going on in the field of CAD and therefore did a feasibility study to see whether this was the right approach for r.c. detailing. It rapidly became clear that 'raw' CAD, used purely as a draughting aid, was not economical compared with the advantages of a system aimed at one specific task.

16. A decision was made therefore to design a system dedicated to r.c. detailing thus getting the maximum benefit out of the computer at a sensible cost. Great care was taken however to design the system in such a way that advantage could be taken of the latest developments in computer technology. This has proved to be very successful in that our customers regularly update their equipment knowing that BARD will still be available to them.

17. With the advent of lower-cost CAD it is tempting to take advantage of the flexibility built into BARD to link it to some of these systems. The economics of doing so would make sense only if a firm already had a CAD system since the combined cost would otherwise be beyond the reach of most small firms. Even then the end result might be solely 'cosmetic' with little or no cost benefit to the user.

18. The next stage will therefore be to increase the capability of BARD so that a greater percentage of the total structure can be detailed with the system. The intention is to add programs to deal with non-rectangular slabs and staircases in the near future.

19. In conclusion, detailing reinforcement manually is likely to be error prone - particularly when producing the bar-schedule. Computers can certainly help to take the drudgery out of a large percentage of this work but, even with the very latest technology, doing the whole job by computer is questionable. BARD was designed to tackle the straightforward elements, recognising that the remainder might be produced more economically by some other means. In the past this has meant manually, but firms who already have CAD systems may prefer to integrate BARD with CAD. Thus BARD may be considered to compliment CAD rather than being in competition with it.

Paper 2. The design and implementation of computer aided reinforced concrete detailing

T.J. PARSONS, BSc, MICE, FEA Ltd, and S.P.R. VINCENT, MA, MSc, MICE, MIStructE, Scott Wilson Kirkpatrick & Partners

SYNOPSIS. This paper describes the design philosophy and implementation experience gained in the development of the GIPSYS computer aided RC detailing system. The question of 2D or 3D is examined with explanations given for the chosen philosophy of starting with 2D but with 3D aspects, laying the foundation for future expansion into 3D. The structure of the RC data, how the associations between details are handled, and how the integrity of the details is maintained so that the automatic bar bending schedules can be used with confidence is discussed.

BACKGROUND AND DESIGN PHILOSOPHY

1. SWK have had considerable involvement in the development of CAD techniques, and the GIPSYS CAD system, now marketed and developed by FEA, is the result of over 15 years work. Although GIPSYS has wide application in many aspects of civil and structural engineering drawing work the initial interest in developing such a system came from the structural engineering departments where it would be used for general arrangement and reinforced concrete detail drawings.

2. The early methods of producing RC drawings involved digitizing directly from roughly drafted drawings (ref. 1). Automatic bar bending schedules were produced but there were several problems with the accuracy of digitizing and registration with the concrete outline. These early developments however provided a sound base of experience from which a second generation RCD system could be developed. Rapid advances in graphic screen capabilities mean that input from drawings fixed to a digitizer has now become unnecessary.

3. One of the drawbacks of the original system was the limited types of RC details that could be produced, mainly slabs and walls. So it was decided from the very outset that the new RCD system should have no limitation on the type and complexity of detail that it could produce so that users would not be discouraged because they could not complete a detail. This alone proved to be one of most onerous design requirements for the system.

4. One of the major considerations in the design of the system was whether it should be wholly based on a 3D model. Whilst such an approach would give many advantages it was decided that such a concept was only viable where the model was a simple one or where a large amount of time could be spent in defining the model. However in civil/structural engineering structures are rarely straightforward and as they are almost always "one off" the luxury of spending a long time defining, designing and drawing them cannot be afforded. It was therefore decided not to dive headlong into attempting to define a 3D model with 3D RC details.

5. A better approach was considered to be to provide a system which could draw any RC detail required of it from the very simple to the very complex but to design it in such a way that in the future the straightforward elements in an RCD structure such as beams, columns and slabs could be detailed automatically straight from a design module or from some form of parameterized reinforcement arrangement with the complex detail being completed by the standard detailing features of the system.

6. An important consideration in the design of the system was that it should, as closely as possible, reflect the methods and terminology used in traditional detailing methods whilst not overlooking opportunities to modify this approach to take advantage of the computer environment. Considerable effort has therefore been put into making the system an 'RC Detailing System' rather than merely a drafting system which is capable of stitching symbols and text together to resemble RC details.

7. In order for the system to be able to deal with any type of RC detail it had to be able to cope with any shape of bar. The approach therefore was to develop a method for defining a reinforcing bar of any shape which would naturally include all those defined in BS4466 (ref. 2). There are certain bar bending rules to be adhered to in shape definition but these can be applied generally to any shape, such as minimum bend radii, and minimum distance between bends etc. In addition to automatic checks intelligence can be built into the shape code definitions, for example, to warn the detailer if a dimension exceeds some important value. Much of the information used in the system is defined externally in a user modifiable data file, which enables the system to be tailored to the individual requirements of a particular organisation or project.

DATA STRUCTURE AND IMPLEMENTATION

External data

8. Many of the parameters which specify the drafting style are defined in the external data file. Examples of these parameters include the lettering sizes and pen thicknesses for each part of the details, the method of numbering variable bar marks, and the layout and headings to be used

for the bending schedules. Standard parameter values such as concrete covers and lap lengths can also be defined for use in calculations.

9. All of the information about the shape codes used by GIPSYS RCD is also defined externally from the program code. This information starts with tables defining the minimum bend radii, bob lengths, and hook lengths for the steel types and bar diameters which are to be used. The definition of each shape code which follows is sufficient to construct an accurate three dimensional drawing of any bar, although the majority of shapes can actually be defined in two dimensions. Standard geometrical checks are carried out automatically on every bar, and additional checks to detect specific errors or issue standard warnings can easily be added to the shape code definitions. All shape 99's must be fully specified as additional shape codes. This makes sure that all bars including shape 99's are properly checked and accurately drawn.

Storage of details

10. All of the details relating to one drawing are stored in a single file, and the software and data structure are designed to maintain the integrity of this RCD details file under any circumstances. If a power failure occurs, or the computer runs out of disk space, or any other catastrophe occurs, the working session can later be resumed without any danger of the data being corrupted, and without losing the work carried out before the failure.

11. The RCD detail file is divided into two main parts, the bar mark table and the reinforcement details. The bar mark table contains the steel type, bar diameter, shape code and dimensions of every bar mark which appears on the drawing.

12. The reinforcement details are subdivided into groups, where each group represents a structural element, such as a beam or a bay of slab. Each group appears separately on the bending schedule. Groups may be repeated or multiplied for similar bays of slab or identical beams or other elements.

13. Each bar may be drawn in as many places as the details require, for instance a bar may occur in several sections as well as the main plan or elevation. All of the drawings of a single bar are associated together, so that it is possible to check that every drawing of a bar is associated with a callup on the finished drawing.

14. Regularly repeated bars may be indicated using ladders. A ladder may be straight or radial, and can appear in a variety of forms for different types of detail. Ladders can also be used to indicate ranges of variable bar marks (Figs. 1-2).

15. Every bar on the completed drawing must be associated with a callup. The callup always indicates the number of bars, the steel type and diameter of the bar, and the corresponding bar mark. It may also contain the pitch of a

fixed pitch ladder, position information such as reinforcement layer numbers, and any further annotation which the detailer wishes to add (Fig. 2).

Detailing method

16. Before any detailing can start, a suitable background drawing on which the details can be constructed must be prepared using the standard GIPSYS drafting system (ref. 3). Each element of the drawing is constructed accurately in full size co-ordinates, so that the dimensions of bars can later be measured directly from the background. The plans, elevations, and sections which form the drawing are then assembled at appropriate scales and positions using the GIPREG drawing register system. Parts of plans used at one scale for general arrangement drawings may be windowed and used at different scales for reinforcement drawings.

17. All standard BS4466 shape codes are available on the digitizer menu used to issue commands, while non-standard shape 99 codes are typed in at the keyboard. The dimensions of each bar mark can either be entered directly as numeric values, or measured as distances from the background drawing. Whenever values are input using GIPSYS RCD, they can be modified using the standard parameter values defined in the external data, such as cover dimensions or lap lengths.

18. The input dimensions are automatically checked as specified in the external shape code definitions, and any warnings or errors are reported before proceeding any further. A three dimensional model of the bar is then constructed, according to the shape code definition. Further checks of minimum bend radius and minimum straight lengths are carried out during this process.

19. The bar is then displayed "floating" on the workstation display, and can be mirrored and rotated in three dimensions to obtain the required elevation of the bar.

20. Blobs and crosses are added automatically to indicate changes of direction which are not visible, according to standard detailing practice. All bends and lengths are drawn accurately to scale, and the bar can be drawn as a double line showing its actual width for fitting together complex details. Bars can either be positioned accurately relative to the background outlines, or they can be floated and placed in approximate positions.

21. Where bars are regularly repeated, as for slab reinforcement or links in a beam, ladders are added to indicate the position and extent of the repetition. Ladders may include several extents of bars with different pitches and changes of direction if required. These features allow a single ladder to indicate bars in several bays of a slab, or different spacings of links in a single beam, or to follow an upstand around the edges of a hole.

22. Sections are constructed by adding further drawings of the appropriate bars to the section outlines. Bar mark annotation can then be added to identify bars drawn in

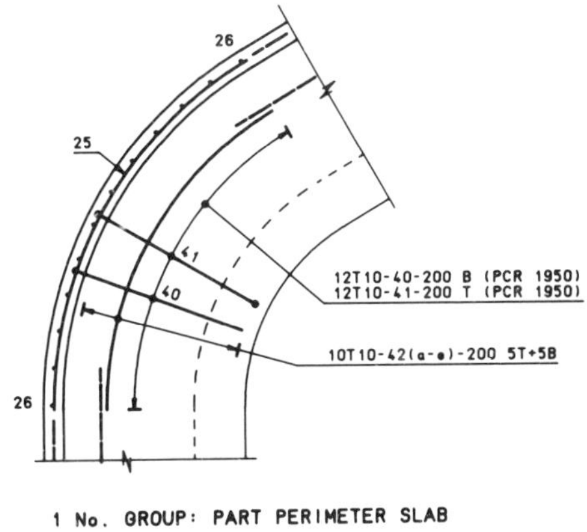

Fig. 1. Radial bars and ladders

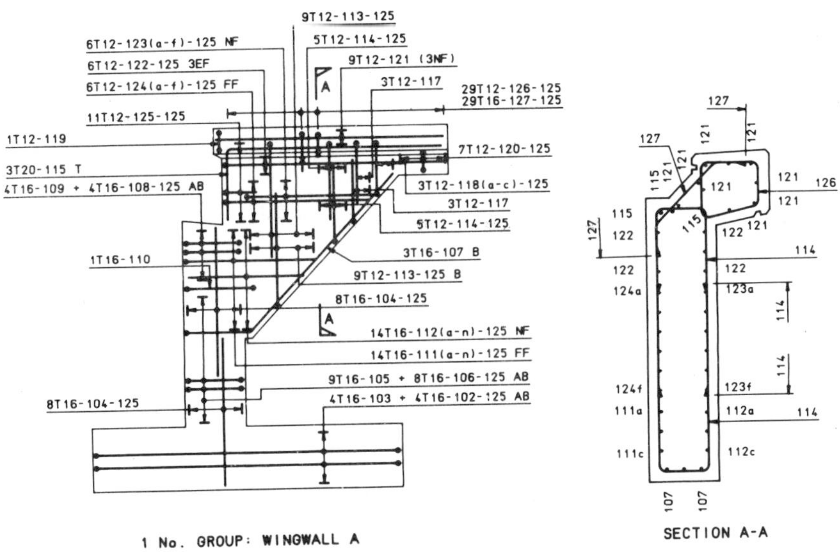

Fig. 2. Typical complex detail produced by GIPSYS RCD

section or the end of bars which overlap, as in Fig. 2. This annotation is associated with the bar itself, and is automatically deleted if the bar is later deleted.

23. The basic information contained in the callup for a bar is generated automatically, while extra information can be added to indicate bar positions. For certain types of position information, the total number of bars will be multiplied according to standard conventions, for instance

the total will be doubled if the same bar occurs in two layers. Several special types of callup are available in association with ladders, including variable bar mark callups and alternate bars along a ladder.

Copy files

24. One very powerful feature is the use of GIPSYS RCD copy files to copy reinforcement details. Any details, from a single bar to a fully detailed group complete with its callups and sections may be copied out to an RCD copy file. This copy file contains a copy of the bar mark information for every bar copied, as well as the visible details. The file may then be copied back either to a different part of the same drawing, or to a completely different drawing. The details may be mirrored or rotated to any angle before being placed back on the drawing, and the software will automatically rotate and shift each item of text to maintain meaningful details.

25. When copying between drawings, the details of every bar mark in the copy file are automatically checked against the corresponding bar mark in the detail drawing bar mark table before the details are added. Alternatively, at the detailer's request, the computer will automatically renumber every bar mark in the copy file to a new bar mark in the bar mark table, and then every occurrence of each bar mark in the callups and annotation will be renumbered to match.

Editing details

26. An important practical requirement for computerised reinforced concrete detailing is that comprehensive and straightforward editing facilities must be available. If the details can easily be revised at the computer terminal, it is possible to carry out the majority of the detailing process at the computer terminal rather than just tracing completed draft details onto the computer.

27. GIPSYS RCD includes comprehensive editing facilities which are readily available at all times, even during the initial input of details. Whenever details are edited, the software ensures that details cannot become inconsistent or incomplete as a result of the editing operation.

Bending schedule

28. The bending schedule is extracted directly from the callups which are visible on the completed drawing, and hence it is always consistent with the plotted drawing. As the computer searches the details for callups, it also checks that every drawing of a bar is associated with a callup.

29. Bar marks are totalled and sorted into order for each group. The headings and notes which appear on the schedule can be adjusted using external data, and the schedule can either be printed on standard computer forms, or on A4 pages in a format based on the layout specified in BS4466.

PRODUCTION EXPERIENCE

30. Whilst every attempt was made at the design stage of the system to cover all requirements it was felt inevitable that modifications and further developments would be necessary in the light of user experience. The modular design of the system allowed ideas to be modified and enhanced as it was developed.

31. One of the most difficult problems to solve has been that of permitting some form of draftmans' licence but at the same time maintaining the integrity of the bar bending schedule. In certain cases it is impossible to draw reinforcement bars to the correct scale and/or in the correct position and maintain clarity.

32. The dimension and accuracy of the bars is always maintained because the dimensions are calculated and stored in a table of bar marks. Therefore it doesn't matter what view of the bar is drawn as the dimensions for scheduling still remain correct. Although it is intended to include facilities in the system to adjust the drawn size of the bars the desired result has been achieved by the users' by simply rotating the bar in three dimensions so that the projected view is the right size.

33. Another problem was that of positioning the tags marking the ladder extents. This was solved by first accurately defining the ladder extents from which the number of bars/pitch calculation is done and then allowing the ends to be moved by a maximum specified amount but without altering the number of bars. The amount by which the extent can be edited is controlled by a factor held in the external data file.

34. The original design of the system required a bar to be placed on the drawing and then the callup added immediately afterwards. User experience showed that it was necessary to be able to lay bars out and then add callups later. This generated the need for a facility to tell the user which bars did not have callups associated with them.

35. Editing can often be quite extensive on a RC drawing. Initially a bar had to be deleted and redrawn if a dimension or the diameter changed, the idea being that the drawn dimensions, especially the bend radii and bob lengths would remain correct. However for small dimension changes this became irritating and time consuming especially as the detailer felt that the drawn view of the bar was adequate. To overcome this problem facilities were added to allow a bar dimension to be changed without requiring a redraw. The amount by which the dimension can be changed is controlled by a factor in the external data file. If so desired redraws can be made compulsory by making the allowable amount of alteration zero. A dimension change only affects the schedule, however a bar diameter change affects the callups on the drawing so facilities had to be added so that if a bar's diameter was changed then all the callups would be automatically changed.

36. As the bar bending schedule is a valuable 'free' product of the detailing process every effort has been made to maintain the integrity of the schedule. Ideally it should only be necessary to check the drawing and take the schedule as being correct as it has been 'automatically' produced. In practice this is not entirely true. Dimensional checks on bar marks are required as the detailer could have chosen a view of a bar such as to 'hide' its true size. However numbers of members and numbers of bars can be relied on.

37. Also in an effort to prevent 'corruption' the callup annotation has had to be strictly controlled so that only a fixed set of bar placing texts can be used. For example 'EF' implies twice as many bars so the system needs to know this to correctly calculate the number of bars. The detailer is allowed to add further annotation to a callup, however it is deliberately enclosed in brackets so that the checker knows that the system cannot have checked for any implication on the calculated number of bars.

FUTURE DEVELOPMENTS

38. As already stated it was a deliberate decision not to launch immediately into 3D, however certain design features have allowed for further development into 3D. All bars are three dimensionally defined and by the addition of a third placement dimension it will be possible to create automatic cross-sections.

39. It is already possible in the system to copy arrangements of bars within a drawing and from drawing to drawing and it would be of great benefit to have parameterized arrangements of bars to enable speedier detailing of the simpler elements. The ability to add bar arrangements created elsewhere to the drawing currently being detailed, with full checking of bar marks to prevent duplication, has enabled links with design to be considered. It is envisaged that many details will be generated by design programs and added directly to the drawing with any complex detail being completed by the standard facilities within the system. This will no doubt lead one day to the engineer being able to define a complex structure in 3D with the design and detailing being automatically carried through to the construction drawings but very importantly with the engineer having full control over all stages of the process.

REFERENCES

1. WILLIAMS G.M.J. and PARSONS T.J. A computer aided drafting system for detailing reinforced concrete. The Structural Engineer, November 1980, Volume 58A, No. 11.

2. BS4466. Specification for bending dimensions and scheduling of reinforcement for concrete, London, British Standard Institution 1981.

3. PARSONS T.J. GIPSYS - a CAD system developed by a user. The Structural Engineer, January 1985, Volume 63A, No. 1.

Paper 3. Reinforced concrete detailing and 3D modelling

M. RICHARDS, Laing Technology Group

SYNOPSIS. This paper outlines the uses to which CAD has been put within the Laing organisation. It takes one specific area of application, RC Detailing and describes in detail the capabilities of the system as developed by Laing and Computervision Inc.

1 In 1975 John Laing purchased a Computervision CADDS3 computer aided design and drafting system to carry out investigations into the use of CAD in the construction engineering and building environment.

2 The area of investigation was computer aided design and pricing, the emphasis to be on the design, in architectural and structural terms, and the initial and final pricing of that design. The initial investigations showed that we needed to gain experience for the type of application envisaged. The system was used successfully as a design and drafting tool with real cost viability.

3 In 1986, because of the lack of the right type of application software Laing decided to develop their own, based on past experience of some twenty years with computers and eleven years of using CAD.

4 The developments being carried out jointly by Laing Design and Elstree Computing Ltd., the software development company within the Laing Group, are:-

a) 3D MODELLING AND REINFORCED CONCRETE DETAILING.

b) DIGITAL GROUND MODELLING.

c) HIGHWAY DESIGN.

d) SITE ENGINEERING.

e) DRAINAGE.

CAD SYSTEMS

5 The reinforced concrete detailing software was first developed in 1976 as a 2.1/2 D detailing system. The current system is based on our experiences of using the original version and enhancing it to a full 3D version. The basis of the system is a 3D structural modelling system with an intelligent data-base.

6 Automatic checks built into the system allows for:

a) Concrete covers.

b) National Code requirements (tension and compression laps etc)
c) Bending and cutting tolerances.
d) Automatic scheduling.
e) Automatic weighting.

7 The system is fully flexible and not restricted to standard elements, beams, columns, slabs but allows for any 3D geometrical shape that can be defined within the modelling system.

8 Future development of the system will allow the design process to be carried out from architectural conception through interactive structural analysis and design to the final automatic or semi automatic detailing process.

R.C.DETAILING

1. The use of CAD within the ABC (Architect, Building, Civil) industry has taken,and is taking many years to really make an impact upon the way in which we work and think.

2. Computers, generally in the Building and Civil Engineering activities have been used for many years but have been limited to the large analytical number crunching activities of the management and commercial need, of those organisations

3. Truly interactive computing allowing the designer a single tool to express and analyse the design requirements of any project and to produce as an end product the 2D communication capability, drawings and calculation, have not been fully adopted.

4. The activities of design have always and still are seen as discrete activities, computers or calculators to carry out computation, pen and paper to labourously write down the results and to explain the method of calculation and pencils and drawings to communicate to the construction site the actual building requirements.

5 Reinforced concrete detailing and the production of bending schedule is one such discrete activity that has been singled out for development using interactive computing methods.

6. To develop an electronic method of reproducing R.C. drawings has to the larger part only resulted in the development of draughting systems and has not taken into account the whole procedure that enables the R.C. drawing to be produced.

7. The R.C. detailing drawing is the end product in a chain of activities of design, covering the Architectural concept, the structural analysis and design and the positioning of space and services to produce the required building.

8. In its own right the RC detail is dependent upon the method of analysis, and the design process has to conform to a set of rules and regulations governed by National Codes and good engineering practice. At the end of this process, the RC detail has to be checked to see if it conforms to the codes of practice and has been interpreted correctly from the theoretical design requirements.

9. To allow all of these requirements to be carried out we must develop systems that can check, backward and forward, the detailing activity so that a better product is produced and is cheaper and more efficient in overall terms. If we cannot, then one should stick to the drawing board and normal methods, as this will be found to be cheaper than carrying out the same manual activity, with expensive computing equipment.

10. To try to satisfy the global use of computing as its relates to RC detailing Laing Design and Elstree Computing Ltd joint venture with Computervision Inc have developed a system that encompasses the majority of the above requirements.

11. The system has been developed, to increase the power of the drawing requirements such, that the calculations of bar lengths, the weights, scheduling and checking against relevant codes of practice is automatic. Apart from the time savings achieved in the drawing requirement, the necessity to check schedules against drawings has been completely removed. The checking procedure normally accounts for some 30% of the over all drawing production time and is a significent cost saving.

12. This has been achieved, by using the Computervision 3D modelling system SMS as the basis for generating the 3D structural elements of the items to be detailed.

CAD SYSTEMS

THE STRUCTURAL MODEL

1. The structure modelling system is based on elements representing the structural members:-

(1) Cols
(2) Beams
(3) Slabs
(4) Walls

2. Graphically and mathematical these are by closed polygonal strings representing the bounding surfaces of each element.

3. In the normal situation, a beam for instance, is seen as 6 No surfaces. The system however is not limited to rectilinier shapes but can be defined as prisms whose cross sections can be any polygonal shape.

4. For ease of insertion, we insert elements in the form of beams, slabs, columns, holes etc.

5. The elements may be positioned as 3D elements in 3D spaces by use of vector and rotational commands.

6. Junctions between the elements, beam, columns, slab junctions are automatically cleaned up by the system,to represent graphically a true 3D solid object not a series of 2D elements viewed from different angles. In addition hidden line removal from the view position is an automatic function.

7. The surface of the elements may be picked up, by digitising, and are used as the reference surfaces when positioning reinforcement within the elements.

8. At present the system starts by manually inserting the elements into their correct 3D space,to represent a building or structure.

9. This process could be carried out automatically as part of a frame analysis and design process, but further work has yet to be completed to enable this to be a reality.

10. The process itself is in fact quite feasible, if we use some of the currently available computing method already on the market.

11. For instance, instead of inserting the beam, slab and column elements, it would be possible to derive the main structural model from the conceptual architectural model.

12. If we take a frame analysis program such as Sstress, we can define a wire model representing the structural

elements and generate this graphically. Onto this wire frame, the designer could impose the loading requirements, floor loading, wall loadings, point loads or impose the requirements for live load and wind loads etc.,

13. These loads could be collated with reference to the designers requirements and used as input to the analysis programes. This automatic collation of loadings would be far faster than typing in loading requirements to an analysis programe, as is normal in the majority of cases.

14. The designer could guide the system to carry out the type of analysis required in terms of full frame analysis, continuous beams, one way or two way spanning slabs etc.,

15. The results of the analysis can be graphically represented on the wire frame model,in the form of bending moment diagram and shear force diagram. From this analysis the designer could then use the system to design the member sizess and output the reinforcement requirements.

16. The sizes of each member could then be applied to the wire frame model and the true 3D object produced automatically with very little further input by the designer.

17. The reinforcement could be inserted automatically with additional construction reinforcement added by the user.

18. This situation has not yet been developed but with the new modeling capabilities it will not be long.

DRAWING OUTLINES

1. For the purposes of drawing productions each element may be displayed individually as required for normal drawing practices. As a plan for slabs or in elevations for items such as beams.

2. Sections through the element to be detailed are generated at any position as indicated by the user automatically. Section marks on the plan and elevations are positioned automatically,as are the section numbers under each section generated. The sections, are not individual drawings,but are in fact view points on the 3 dimensional model, such that any change to the model is automatically reflected in the sections.

3. We have already mentioned the capability of automatic line removal, in addition, within the detailing package, hidden lines are automatically generated with a dashed line to represent items on the far side of the main elements. For

plotting purposes the the elements and sections are automatically generated on separate layers such,that different pen thickness can be used - say .2 for hidden lines .4 for lines in elevations and .5 or .6 for the lines representing the sectioned elements.

R.C.DETAILING

1. The reinforced detailing system is based upon three external parameter files that can be user defined.

(a) Code of Practice Parameters.
(b) Project Specification.
(c) Bar shape Parameter File.

Code of practice parameter file.

2. The code of practice parameter file allows the user to define these items with any particular code of practice,that he wishes to be checked during the detailing process.

(a) Bar Diameter.
(b) Type of Bar (High Tensile Mild Steel)
(c) Minimum Spacings.
(d) Bar Bending Requirements.
(e) Minimum bar lengths.
(f) Dimensional Tolerances
(Bending Tolerances 4)
(g) Tension Laps Compression Laps etc.,
Cutting Tolerances.

3. This is not a full list, but an indication of the types of requirements.

4. The Parameter file has been made user definable, to allow the system to be used with any national code rather than hard coded to one.

5. Even in the UK we detail to different National codes depending on the areas in the world in which we have contracts. Bar diameter for instance, may be specified in real terms for positioning requirements, but output in nominal terms for scheduling purposes.

6. It also allows for national requirements as used by the USA (# 3) for call ups by using the actual diameter for calculation purposes.

7. A full list of tension and compression laps can be stated, depending on steel strength, concrete strengths, and for laps to be used for special requirements - water retaining structures and for aggressive environments.

```
3!******************************************************4!*
5!*                  CODE OF PRACTICE PARAMETERS
6!*
7!*
8!*
9!*    Note!!!  Only Parameter Values may be altered.
10!*
11!**********
12!*
13!****************************************************14!* No
Type          Value      Despription
15!****************************************************
16!   01     T          B          Standard (B,G,A,F,)
17!   02     F         40.0        Minimum Steel Cover.
18!   03     F         50.0        Minimum Bar Spacing.
19!   04     F       1000.0        Maximum Bar Spacing.
20!*
21!***     Allowable Steel Types
22!*
23!  100     T          R          Steel Type No 1 (Grade
                                 250 Plain Bars)
24!  101     T          T          Steel Type No 2 (Grade
                                 460 Deformed Bars)
25!  102     T          0          Steel Type No 3 (Not
                                 a Valid Steel Type)
26!*
27!*** Hook allowance Table for Steel Type No. 1.
28!*
29!  110     I,F        6,100.0
30!  111     I,F        8,100.0
31!  112     I,F       10,100.0
32!  113     I,F       12,110.0
33!  114     I,F       16,150.0
34!  115     1,F       20,180.0
35!  116     I,F       25,230.0
36!  117     I,F       32,290.0
37!  118     I.F       40,360.0
38!  119     I,F       50,450.0
39!  120     I,F        0,0.000
40!  121     I,F        0,0.000
41!  122     I,F        0,0.000
42!  123     I,F        0,0.000
43!  124     I,F        0,0.000
44!  125     I,F        0,0.000
45!*
46!***   Bend Allowance Table for steel Type No 1.
47!*
48!* 130     I,F        6,100.0
49!  131     I,F        8,100.0
50!  132     I,F       10,100.0
51!  133     I,F       12,100.0
52!  134     I,F       16,100.0
53!  135     I,F       20,100.0
```

CONTRACT PARAMETERS

```
7!*****************
8!*
9!*    Note  !!! Only parameter values may be altered.
10!*
11!****************
12!*
13!*********************************************************
14!*   NO         Type        Value      Description
15!*********************************************************
16!*
17!***       Layer Allocation
18!***       (-1 Denotes active construction layer)


121!101      T    M           * Data base bnits
122!102      T    MM          * User communication units
123!103      F    0.001       * Unit conversion factor
124!*  N.B. All numbers input by the user will be
        multipled by this factor.
125!*

126!104      F    0.025       * Length rounding constant
                                in DB Units.
127!105      F    0.005       * Bar parameter rounding
                                constant in DB units.
128!106      I    2           * No of bar diameters for
                                internal radius
129!*
130!107      I    3           * No of bar diameters for
                                internal radius
131!*                         * (Steel type No 2
                                <= parameter 205)
132!108      I    4           * No of bar diameters for
                                internal radius
133!*                         * (Steel type No. 2>
                                parameter 205
134!*
135!109      F    2.0         * Default bar label text
                                height (DB Units)
136!110      F    6.0         * Bar extent lines
                                (DB Units)
137!111      F    1.75        * Bar extent arrow size
                                (DB Units)
138!112      F    1.25        * Bar extent square symbol
                                size (DB Units)
139!*
140!115      F    0.025       * Bar space rounding
                                constant in (DB units)
141!116      F    12000.0     * Maximum Bar length
142!*
143!***      Table of nominal to actual bar diameters 144!***
for steel type No 2 - (Grade 468 deformed bars <T>
```

PROJECT SPECIFICATION

1. An additional file has been allowed for, where specific Project specifications require that the National Code in certain areas is to be ignored or overridden.

BAR SHAPE PARAMETER FILES

1. This file is possibly the most important file, as it has control over the methods of insertion and methods of calculation for any given bar shape.

2. Basically it is a symbol file defining a shape but the elements of which are variable parameters. For a shape 37, we know that it consists of two legs at right angles to each other, but the lengths are totally variable depending, on their position within the structure.

3. The shapes are bases on a series of 3D nodes. The nodes represent the start and end of the bar and each bend, for a shape 20 2 nodes are required, shape 37 3nodes are required and so on for any shape required by the user. This means that special bar may be generated by the user and not restricted to the prefered shapes.

4. An example of a bar parameter file is shown below.

5. Standard hooks and bobs are defined within the system and can be positioned with the parameter file either at the start or end of a bar

Code:	represents the shape code used at the time of insertion and can be any user defined code.
Report:	allows the user to define the actual shape code to be used on the schedule as well as the length of the individual legs.
Pars:	represents the individual parameters.
Bar:	indicates the graphical representation of each bar shape.
Locs:	indicates the number of nodes and digitises requires to define and position the bar within the element.

```
5************************************************************ 6*
7!*  Bar Shape Code Definition file.
8!*
9!************************************************************
10!*
11!*
12!     Code        :20
13!     Pars        :A
14!     Locs        :P1,P2
15!     Bar         :(), (X+A,)
16!     Set         :A=$x(P2-P1)-00001
17!     Len         :A
18!     Report      :20
19!     End
20!***
21!     Code        :32
22!     Pars        :A,H
23!     Locs        :P1,P2
24!     Bar         :(), (X+A,)
25!     Hook        :(X+A-D1-R,>,<X+R+(D1/2),),
                   (R+D1/2), (,Y+(2*R)+D1),
26!     $:           (R+D1/2), (X-(H+2*R+1. 5*D1 -
                   (P1* (R+D1/2>>>>
27!     Set         :A=$X(P2-P1+(D1/2)- 00001
28!     Len         :A+H
29!     Report      :32,A
30      End
31!***
32!     Code        :33
33!     Pars        :A,H
34!     Locs        :P1,P2
35      Hook        :(X+(H+2*R+1,5*D1-
                   <P1*(R+D1/2))), Y+(2*R)+D!),
36!     $:           (X-(H+2*R+1,5*D1 - (P1*(R+D1/2))))
37!     $:           (R+D1/2, <Y,-2*R-D1>,(R+D1/2),
                   (X+D1/2+R)
38!     Bar         :(), (X+A-D1)
39      Hook        :(X+A-D1-(R=D1/2)), (X+R+D1/2>,
                   (R+D1/2).
40!     $:           (,Y+(2*R)+D1), (R+D1/2),
41!     $:           (X-(H+2*R+1, 5*D1-(P1*(R+D1/2>>>
42      Set         :A=$X(P2-P1)+D1-00001
43!     Len         :A+(2*H)
44!     Report      :33,A
45!     End.
```

6. The flexibility of the parameter file is such, that a specific bar may be calculated in a number of different ways. A shape 41, can be specified and reported in the normal fashion. It is possible however, to define a shape 41, as is normal for graphical representation, but to be reported as a shape code 20 when say the bar offsets are less than the bar deformation capabilities say for a T12 diameter bar.

METHOD

7. In a paper and presentation of this length, it would be impossible to describe every command available within the programe. Instead we will limit the paper to the insertion and checking of bars within the structural elements.

8. To insert a bar, the command structure consists of a VERB, a NOUN, some modifiers and instream commands of information called GETDATA.

The command is-

INS BAR the system then requires information about the bar itself.

(1) Steeltype.
(2) Bar Dia.
(3) Number of Bars.
(4) Or spacing of Bars.
(5) Shape code.
(6) Steel layer.

This allows the bar position to be indicated with reference to the concrete surface.

(6.1) T1
(6.2) T2
(6.3) B1
(6.4) B2 etc.,

The full command system would be:-

INS BAR STYPE T20 SPACING 250 SLAYER T1 SHAPECODE 37

9. After the colon has been typed the system shapecode 37 requires surfaces to be digitised to give the 3D model positions of each of the bends, plus the two ends of the bar. For the shape 37, this could consist of a minimum 6 digitises, where the bar is parallel to one surface. Or a maximum of 9 digitises where the bar is positioned 3 dimensionaly within the elements.

10 To position the node, into an end corner of a beam for example, all three surfaces, bounding that corner must be digitised. The offset from any surface is taken from the defaults covers laid out in the parameter file or may be entered prior to digitising where offsets are in excess of the defaults.

11 All bar leg lengths will then be calculated from the digitised positions. If specific dimensions are required for any bar leg then they can be entered explicitly as modifier A.B.C.D.etc., In this situation the two consecutive digitised nodes will indicate a vector of each bar dimension and self calculating where no dimensions are given.

12 For a shape 37 A or B or both dimension may be explicit.

13 After positioning the typical bar within the element the user is then prompted by the system to digitise the position of the first and last bars. This may be carried out by digitising surfaces as described for positioning the main bar. The extents of the bar can be locked horizontally, vertically or perpendicular to the first position such that the first and last bar is parallel to the typical bar.

ANNOTATION

1. The bars are annotated and flagged to produce the final drawing requirement as guided by the user. Typical bar extents are fully annotated whilst every other annotation of that bar when it occurs in a section or elevation will be by bar mark number only.

2. Bar marks are automatically inserted and registered such, that bar marks are not duplicated unless specified by the user. If bar marks are to be duplicated then the command would be:-

INS BAR BMARK 20:

All other information will then be inserted by the system, as for the original bar mark. The user will however,have to position the nodes of the typical bar and the extents so that differences in bar numbers can be calculated.

3 Any difference in the location of the bar nodes, leading to differing dimension for any of the bar legs, will automatically produce a new bar mark number, even if a bar mark is specifically stated.

ADDITIONAL MODIFIERS

1. If it is required that one bar should lap with another, then the system need only be prompted for the correct type of lap, prior to digitising the bar to be lapped with. Two type of laps entered are for:

(a) Tlap - Tension Lap
(b) Clap - Compression Lap.

2. The correct tension lap, will then be extracted from the parameter file, for the particular diameter and steel type.

3 The lap are automatically chosen, after digitising the lapping bar; as being the smaller diameter of the two bars, either the bar being inserted or the bar to be lapped with.

4. For positioning bars with reference to bars already inserted the modifiers:-

(a) Ref Bar: with reference to another bar.
(b) Angle: at an angle to another bar or surface.
(c) Perpend: Perpendicular to another bar or surface.
(d) Rotate: About another bar or about itself.

are available to enable the user to manipulate the bar prior to annotation.

CLASH CHECKING

1. During insertion, bars are placed with reference to other bars and so a large amount of checking is eliminated, as it is impossible to make them clash.

2. If bars have not been placed with reference to other bars, that is, when bars have been placed in adjacent slabs, with the same spacing and the same first and last bar position, then it is obvious that they will clash. Global clash detection of this sort would require the system to carry out an enormous amount of computation, every time a bar was inserted, and this would reduce the viability of the system. In these circumstances, it is left to the experienced user to see that these do not occur.

3 There is however, a Status Bar facility, that will give a full report on each bar, given actual verses nominal information, e.g.

Actual Bar Diameter - Normal Bar Diameter.
Actual Bar Spacing - Norman Bar Spacing.

and all of the properties attached to that bar - steel type, number of bars, surface covers etc.

4. The command will report, if a bar is inside or outside of the structure and also if the minimum spacing requirements of any bar given has been overwritten.

REPORT SCHEDULE

1 This command allows bending schedules to be produced to a predefined format.

2 The schedules are produced by element, by drawing, or by a number of elements from a drawing.

3 All schedules indicate the element label in the extreme left hand column.

An example of such a schedule is given in Fig. 1.

4 All schedules can be automatically weighted and weights collated for each drawing or for any number of drawings or for a project.

An example of a Weights Schedule is given in Fig. 2.

AUTOMATIC DESIGN AND DETAILING

1. With this system the simple elements have already been developed to carry out design with automatic detailing and scheduling.

These include:-

(1) Pads.
(2) Pilcap: 1 2 3 4 & 5 pile groups.
(3) Columns.

2. Pads and Pilecaps can be detailed in full 3D automatically or in a tabular form, columns at present are tabular.

Reinforcing Bar Schedule.

project	project no.	drawing no.	schedule	revision
Computer Reinforcement Detailing.	12345	E 1 0 0	01	

drawing title	drawn	checked	date	all dimensions in accordance with BS4466
Beam Drawing.	RD	MR	Oct 1987	

member (mbr)	bar mark	type & size	no. of mbrs.	no. in each	total no.	length of each bar mm	shape code	A mm	B mm	C mm	D mm	E/r mm
Beam on	01	T32	1	4	4	5500	20					
G.L. A	02	T25	1	4	4	5500	20					
	03	T20	1	4	4	1900	20					
	04	T20	1	8	8	1700	37	500				
	05	T16	1	4	4	5500	20					
	06	T25	1	4	4	1900	37	500				
	07	T20	1	8	8	3000	20					
	08	R10	1	56	56	2525	60	780	380			

R – Plain or deformed grade 250 bars complying with the requirements of BS 4449;
T – Type 2 deformed bars of grade 460/425, complying with the requirements of BS 4449 or BS 4461;
X – A general abbreviation for types not covered by R or T, where the properties required have to be defined in the design specification.

Fig. 1. Reinforcing Bar Schedule

REINFORCING BAR WEIGHTS.

project	project no.	drawing no.	schedule	revision
Computer Reinforcement Detailing.	12345	E 100	01	

drawing title	drawn	checked	date	all dimensions in accordance with BS4466
Beam Drawing.	RD	MR	Oct 1987	

Member	Steel Type & Size	Weights in Tonnes.
Beam on	T 16	0.034
G.L A	T 20	0.111
	T 25	0.114
	T 32	0.139
	R 10	0.087

R – Plain or deformed grade 250 bars complying with the requirements of BS 4449;
T – Type 2 deformed bars of grade 460/425, complying with the requirements of BS 4449 or BS 4461;
X – A general abbreviation for types not covered by R or T, where the properties required have to be defined in the design specification.

Fig. 2. Weights Schedule

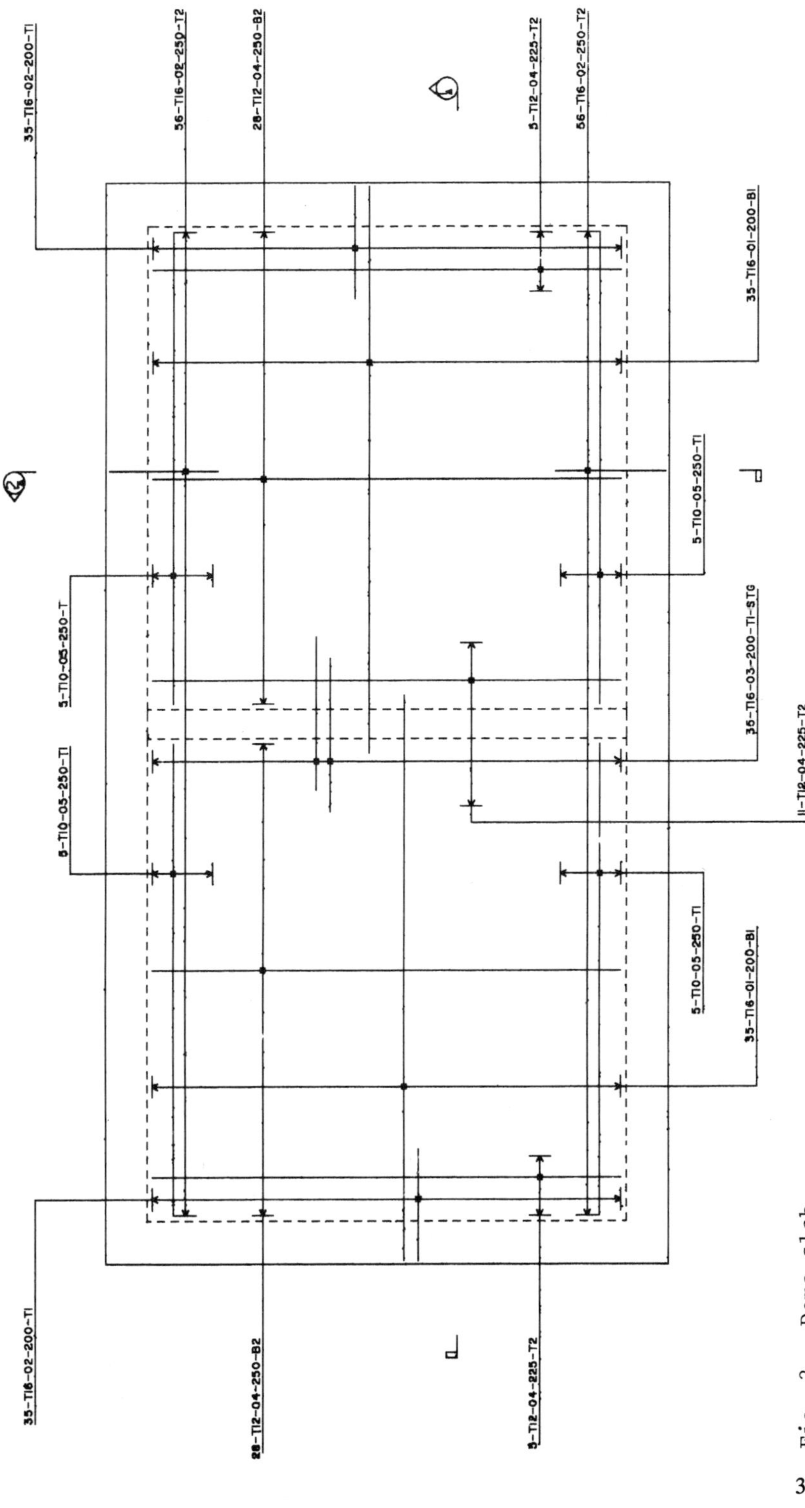

Fig. 3. Demo slab

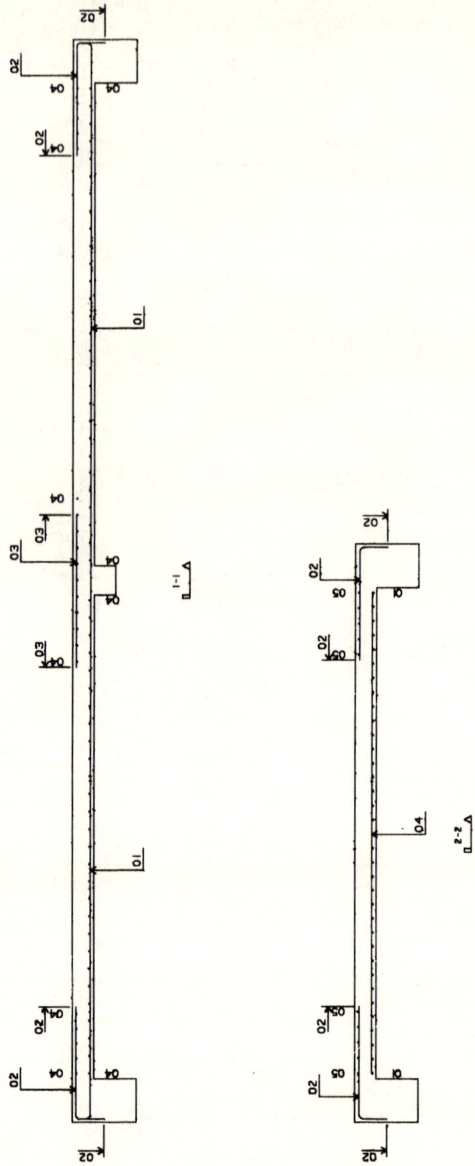

Fig. 4

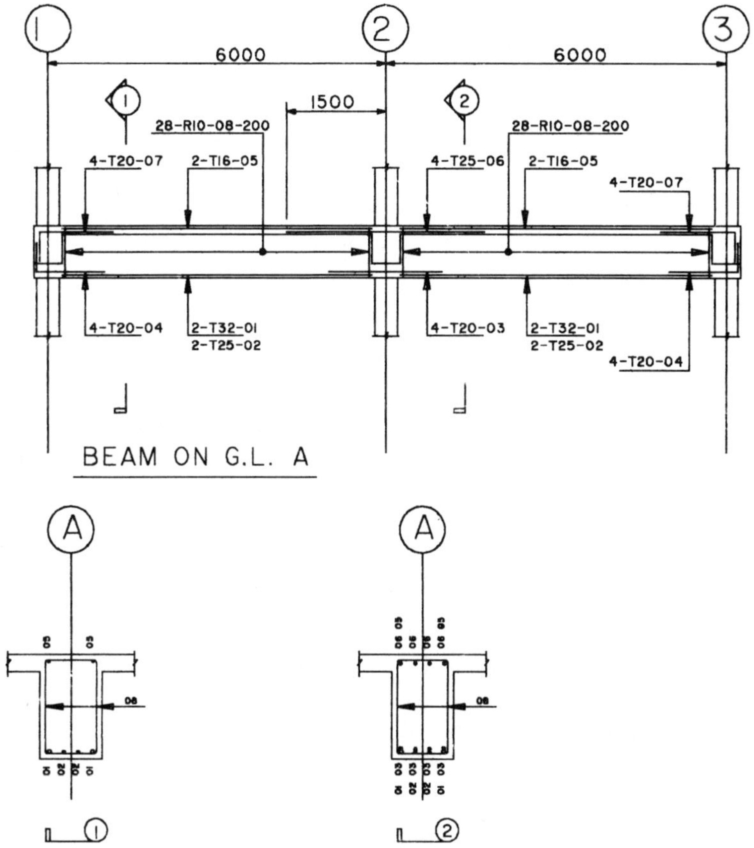
1
2
3
6000
6000
1500
28-R10-08-200
28-R10-08-200
4-T20-07
2-T16-05
4-T25-06
2-T16-05
4-T20-07
4-T20-04
2-T32-01
2-T25-02
4-T20-03
2-T32-01
2-T25-02
4-T20-04
BEAM ON G.L. A
A
A
08
08

Fig. 5

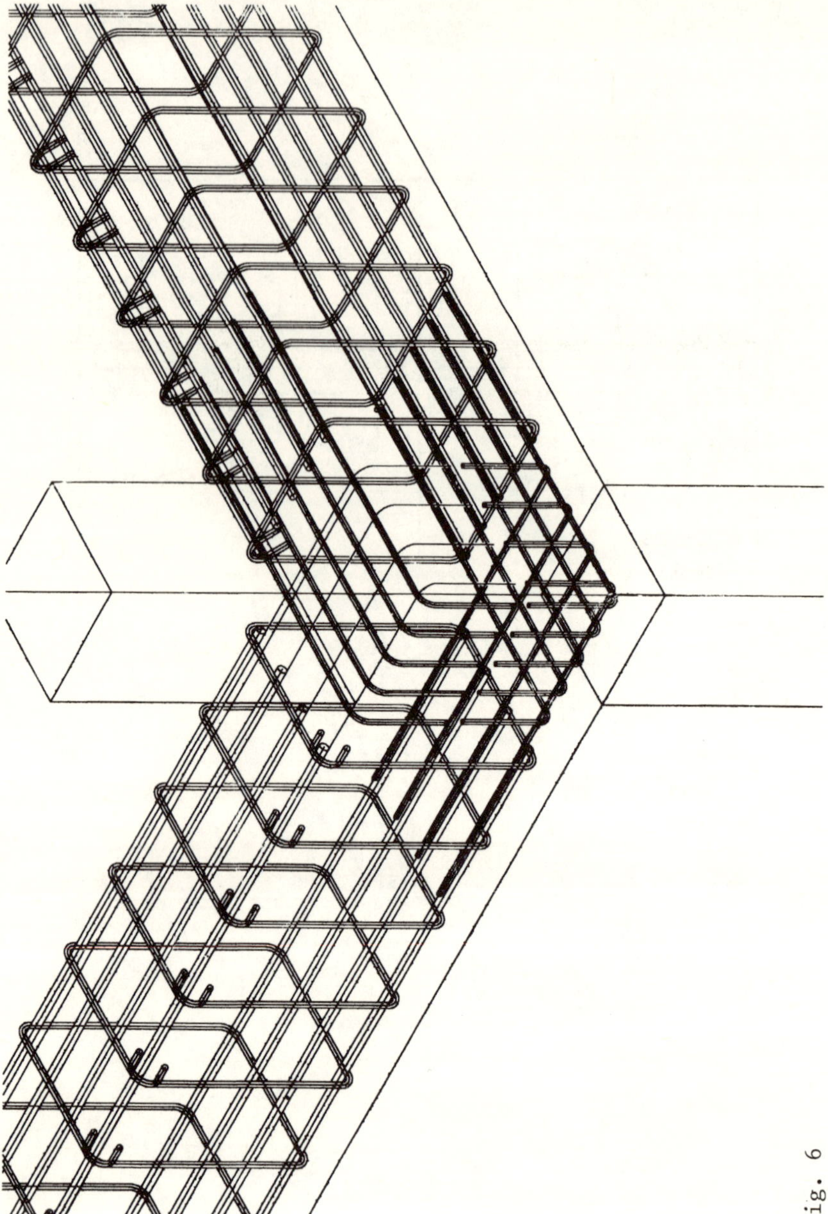

Fig. 6

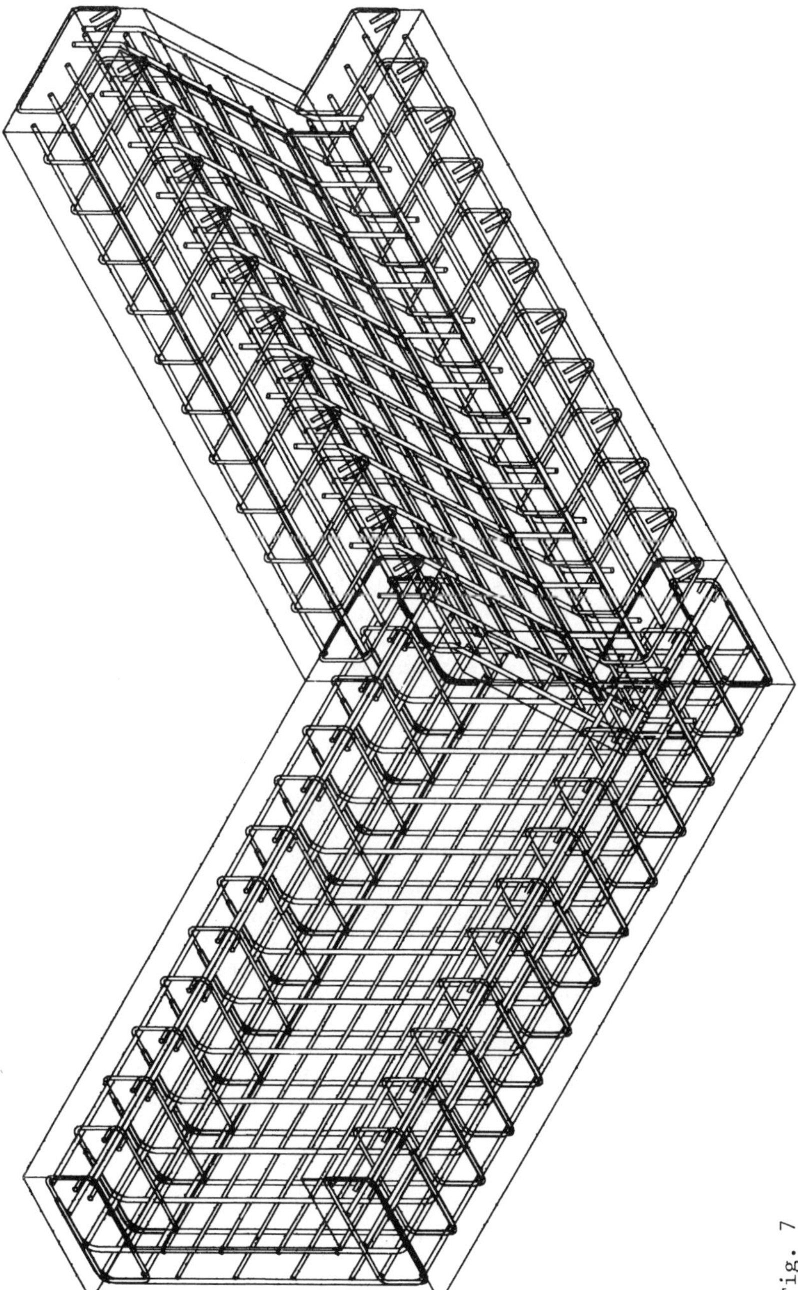

Fig. 7

ADVANTAGES OF 3D

1. Deleting bars from the model,using either,the plan or a section for indicating the bar to be deleted, then that bar will be automatically deleted wherever it is shown in any other section.

2. Additionally when a new bar is inserted into the model,then the bar will be automatically indicated on any section that the bar crosses.

3. This ability obviously has great advantages over normal drawing methods where bars may be omitted from sections due to error.

4. The examples given in Figs 3 - 7 show a single slab with the section generated automatically for the 3D model.

5. The two examples, depicting the full 3D model, with all bars drawn as solid bars in 3D, show the obvious advantages of inspecting simple or complex junctions.

6 This facility must be used with some thought,as the generation of all elements in 3D at the same time can produce, a model that contains so much information that the view becomes a maze of reinforcement. In this situation, it is impossible to see what is happening. This ability is controlled, by the user who, can display individual group of bars that he wishes to investigate, rather than using a global model display capability.

7. Speed of operation of the system, is obviously going to be dependent on the competence of the operator, as a detailer and computer operator. However, under test the system takes approximately 3 minutes per bar group insertion, which for a 200 bar group drawing would give 10 hours of detailing time. As scheduling, weighting and code checking is all part of the operation no additional time is required for those items.

CONCLUSION

1. Reinforced Concrete detailing has at last become a viable proposition, using interaction CAD as a discrete operation. However this should only be regarded as the beginning. The construction and building industry should press the vendor to develop the right type of software, for the application and co-ordinating of all the activities that we require to produce the finished building.

2. At present we seem to accept application software, that sits on as many different systems as there are disciplines, without the ability to transfer information freely between them.

Paper 4. Detailing of reinforced concrete using CAD

J.W. KATTENHORN, Taylor Woodrow Services Limited

SYNOPSIS. This paper considers the approach Taylor Woodrow Construction Ltd has taken in the routine production of reinforced concrete detail drawings using computer aided draughting. The customisation for building design of a proprietary CAD system is described. Conclusions on productivity are drawn and anticipated future developments are outlined.

INTRODUCTION

1. Taylor Woodrow Construction (TWC) is a broad based company having interests in many spheres of engineering activity, although perhaps best known for heavy civil engineering. Involvement ranges from the advanced technologies such as nuclear, the renewable energies and offshore development, to the less glamorous but none-the-less important sector of conventional building structures.

2. Over the years considerable computer resources have been deployed by TWC in 'high-tech' activities but in the 'bread and butter' building areas computing had remained almost totally absent. The introduction of computer aided draughting (CAD) presented an opportunity to redress the balance, particularly in the field of building design where similarity exists between buildings at least as far as the skeletal structure is concerned.

3. An early survey of the draughting undertaken by the Structural Design Department within TWC indicated that reinforced concrete details and schedules represented 60% of total drawing production, the remaining 40% being reinforced concrete and steelwork general arrangements, and other layouts. It was evident that reinforced concrete detailing would be a strategic area of the Department's work to be tackled by CAD.

4. In the late 1960's TWC had, in fact, begun to experiment with automating the design and draughting of r.c. columns. However, the lack of interactive user devices prevented the preparation of drawings to an acceptable standard of presentation. It was necessary to wait a decade for the hardware technology to become commonly available for this type of development to be a viable proposition.

BACKGROUND

5. The multi-disciplinary nature of the Company's business, integrating Architecture, Civil, Structural, Steelwork, Electrical and Mechanical skills, required a general CAD draughting system capable of customisation to the various discipline needs. Evaluation of various systems during 1984 led to the selection of the Computervision (CV) Medusa Drafting System. A four workstation trial configuration based on a Prime Computer, Westward Graphics terminals and an electrostatic plotter was installed early in 1985.

6. Although in TWC Medusa is mainly used for 2D draughting, it has wire frame and 3D solid modelling capabilities, which are useful for architectural visualisation exercises and co-ordination of M & E services. The data management structure within Medusa allows ready customisation.

7. Workstations are located in the engineering departments for use by their own trained users and are not operated on a centralised 'bureau' basis. This ensures a continuing high quality of output commensurate with that produced manually by the departments.

8. The present network of Medusa workstations numbers 21 in 5 separate office locations distributed between the London area and the North of England, as illustrated in the diagram (fig. 1).

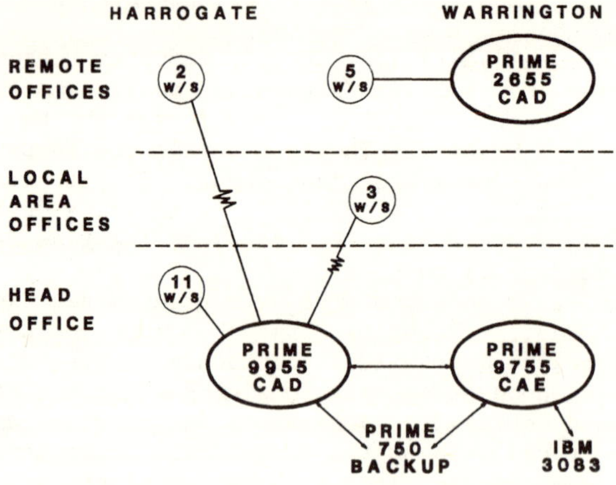

Fig. 1. Workstation network and locations

AEC CUSTOMISATION IN TAYLOR WOODROW

9. Customisation of a CAD system to a company's needs is inevitably expensive but it ensures that the requirements of in-house professionals are fully met. TWC has a long history of software development and possesses the capability to carry out the significant CAD development required for a multi-disciplinary team.

10. The benefits of customisation are 3-fold:-

(a) Draughting using library symbols or programmed graphics generators can be many times faster than is achieved manually or by normal CAD draughting.

(b) Pre-prepared details, symbols and programmed generators conform to Company conventions and not to the individual draughtsmen's differing styles as often tends to be the case with manual draughting.

(c) Drawings produced by a customised system have a consistent internal data structure which ensures that subsequent processes, such as material take-off (MTO), can be successfully carried out irrespective of the originating draughtsman.

11. In this paper we focus attention on a recent customisation for the detailing of reinforced concrete components.

GENERAL R.C. DETAILING

12. CV Medusa market a package for r.c. detail draughting and bar scheduling which provides for detailing based on a bar by bar approach. The detailer selects the type, diameter, shape code and view of a bar from a menu (fig. 2) prior to positioning the bar on the drawing. The bar label is then generated automatically. The user is prompted for the extent and spacing of the bar group or groups. The label is positioned on the drawing independently of the bar, the relationship between bar and label being maintained by the CAD database.

13. A typical bar and label is shown in fig. 3, the bulk of useful data being contained in the bar label. The solid text indicates parameters supplied by the detailer. Hollow text items are calculated by the computer automatically when the drawing is subsequently scheduled. The boxed text is information not shown on a reinforcement drawing (e.g. shape code and limb lengths), but used in the scheduling process. Such information is not normally plotted when a hard copy of the drawing is taken.

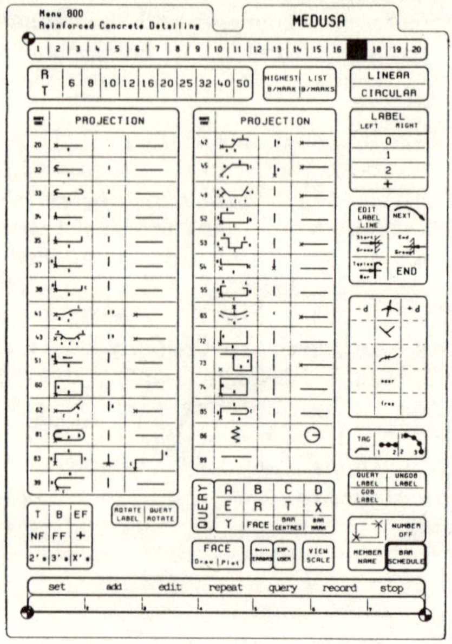

Fig. 2. R.C. Detail Draughting and Bar Scheduling Menu

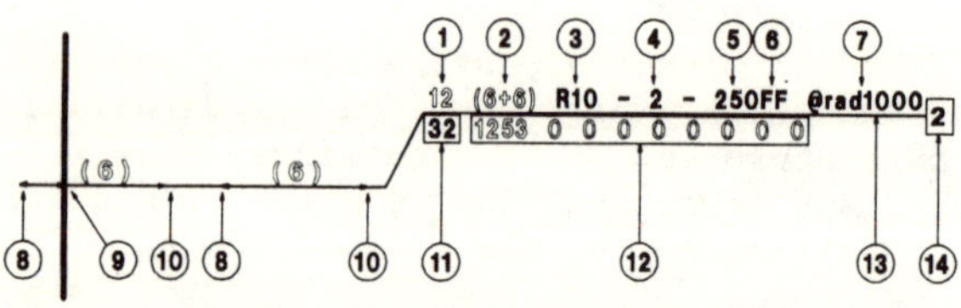

1	Total No. of bars	8	End of bar group
2	No. bars in each group	9	Location of typical bar
3	Material type/diameter	10	Start of bar group
4	Bar mark	11	Shape code
5	Bar centres	12	Bar dimensions
6	Face condition	13	Label line
7	Radius of reinforcement	14	No. of Groups

Fig. 3. Typical bar and label data

14. To produce a schedule it is only necessary for the drawing to be scanned to extract the required data for presentation. Thus integrity between the drawing and the schedule is maintained; if the drawing is correct then so must be the schedule. The scheduling process checks for erroneous duplication of bar marks, with options to merge duplicate bars and/or optimise bar marks.

15. The schedules are produced as separate A4 sized CAD drawings, which allows the detailer to add notes and sketches, (e.g. an illustration of a shape code 99). A typical schedule is shown in fig. 4.

Drawn by PRD Checked by SR Date AUG. '87	TAYLOR WOODROW GROUP BENDING SCHEDULE FOR REINFORCEMENT	Bar Schedule Reference Sht. No. 3335 /1 — Rev.
	CLIENT JOHN LEWIS, PARK ROYAL SECTION GROUND BEAMS Sheet 1 All bar dimensions to B.S. 4466	Ref. Drawing No. TWC/3481/3335

Member	Bar mark	Type and size	No. of mbrs.	No. in each	Total No.	Length of each bar mm	Shape Code	A mm	B mm	C mm	D mm	E/R mm
BEAM B1,B2	1	T25	6	2	12	5800	20					
	2	T25	6	2	12	4000	99	(400) 300 55 2900 450				
	3	T25	6	2	12	5775	37	450				
	4	T25	6	2	12	4675	34	4500				
	5	T25	6	4	24	2000	37	250				
	6	T12	6	2	12	5800	20					
	7	T12	6	22	132	4600	34	4500				
	8	R12	6	21	126	4650	60	2000	200			
	9	R12	6	9	54	3650	60	1500	200			
BEAM B3	5	T25	3	2	6	2000	37	250				
	10	T25	3	2	6	2750	37	250				
	11	T25	3	2	6	3350	38	300	2900			
	12	R12	3	16	48	1475	38	650	225			
	13	T12	3	2	6	1750	34	1650				
	14	T12	3	8	24	2200	35	2000				
	15	T12	3	2	6	1900	20					
	16	T12	3	4	12	2850	20					
	17	T12	3	4	12	1000	20					
	18	T12	3	16	48	1075	35	885				
	19	T12	3	4	12	900	34	800				
	20	T12	3	4	12	1050	35	850				
BEAM B4,B8	31	T20	2	4	8	2300	20					
	32	T12	2	10	20	2300	20					
	33	T20	2	8	16	1450	37	300				

REV.	DATE	DESCRIPTION	BY	CHK.	CHIEF ENG.

Fig. 4. Typical schedule

16. The automatic production of schedules represents a very considerable saving in effort. However our experience is that the overall productivity improvement from using the package is only of the order of 30% over manual draughting whereas for CAD to be cost effective an improvement of at least 50% is required.

R.C. DETAILING FOR BUILDING DESIGN

17. In the context of the Structural Design Department, whilst every building designed is different there is much that is common, particularly when the building is viewed as a collection of components. This provides the key to our approach to the automation of the detailing process.

18. Based on the component concept the principal basic structural components of a building are the columns, foundations, beams, slabs and staircases. A customisation has been undertaken for detailing utilising a suite of parametric structural outlines and reinforcement details.

19. The R.C. Component Detailing package (RCCD) is substantially complete. The core of the RCCD package consists of 8 menu sheets:-

Menu Sheet 1 - GENERAL COMMANDS AND UTILITIES. Using this menu in conjunction with the Medusa R.C. Detailing Package the user can modify standard r.c. components or create non-standard details.

Menu Sheet 2 - FOUNDATIONS. Pile caps, pad foundations and column stubs can be detailed. Circular and universal bearing piles are catered for, in groups of up to 9.

Menu Sheet 3 - COLUMNS. Elevations and sections of rectangular and circular columns can be drawn. Fifteen types of column reinforcement cage, together with six top steel and two bottom steel arrangements, have been defined. Slab and beam details can be simply superimposed onto the column. This menu is reproduced in figure 5.

Menu Sheet 4 - BEAMS. Single and continuous spanning rectangular beams are catered for using loose bar reinforcement arrangements.

Menu Sheet 5 - SLABS. Slabs (hollow rib and uniform thickness) can be detailed.

Menu Sheet 6 - STAIRCASES. Single flight staircases, insitu or precast, can be detailed. Allowance for the architectural finishes can be incorporated in the definition of the structural outline.

Menu Sheets 8 & 9 - SYMBOL MENUS. Fixed symbols, joint details and general notes. These libraries can be conveniently added to by the users.

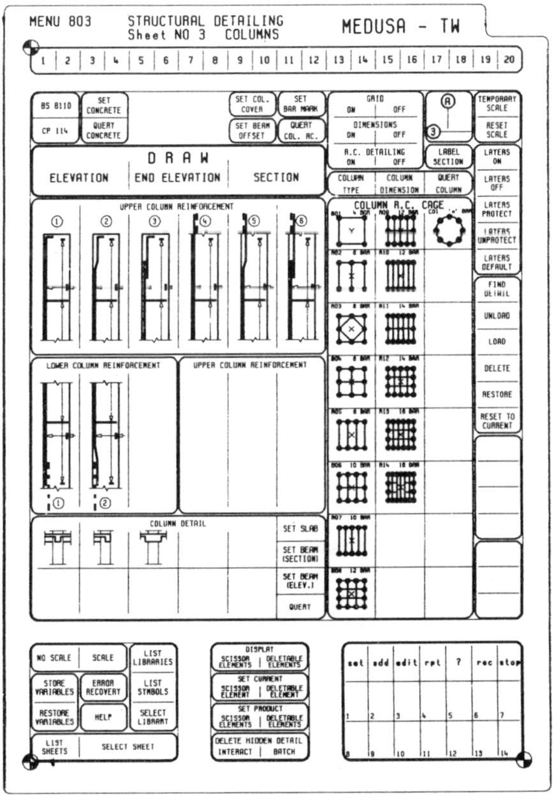

Fig. 5. R.C. Component Detailing Package
Menu sheet 3: Columns

The contents of the menu sheets will be added to as the need arises. For instance, tapered columns and T-beams will be incorporated in the near future.

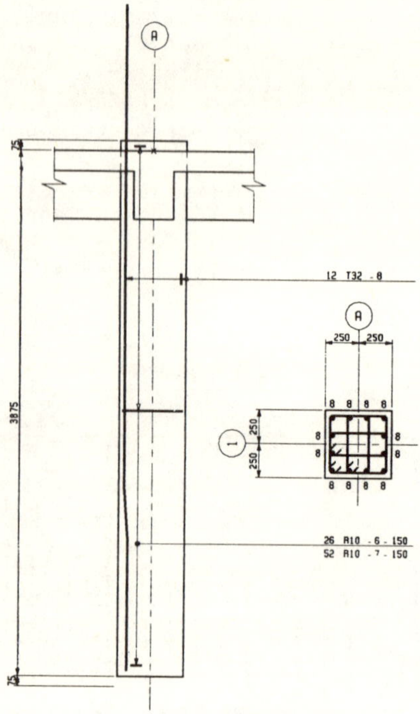

Fig. 6. Typical column detail generated using the RCCD package

20. Examples of the use of the RCCD package are shown in figures 6 and 7. A pile cap or column detail such as those depicted can be generated in less than 45 seconds. The user is only required to enter information on the size of component, concrete cover, bar type and bar diameter where these differ from the user specified defaults. The calculation of bar lap and limb lengths is carried out automatically in accordance with either CP114 or BS8110.

21. The details produced using RCCD conform to the data structure of the CV Medusa general detailing package. The compatibility between packages allows the detailer to develop or edit components to produce non-standard 'specials'. Compatibility also extends to other TWC customisations allowing architecture, steelwork and building services entities to conveniently co-exist on a drawing and to be recognised.

22. Either CP114 or BS8110 codes can be specified for the detailing of components. The principal parameters, such as lap lengths and bend radii, are held in internal tables referenced by the component detailing routines. The routines check that details comply with the relevant code.

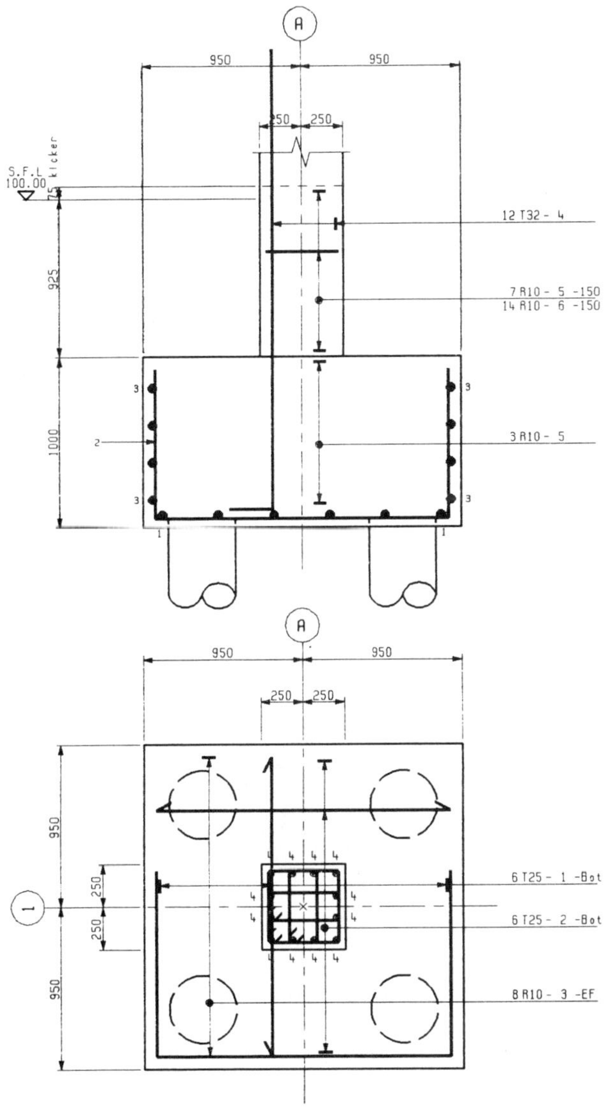

Fig. 7. Typical pile cap generated using the RCCD package

23. The user can enter or modify variables, such as column dimensions, bar type and bar diameter, by either simple menu probes or typed input prior to the drawing of the detail. Once the value of a variable has been entered it becomes the current default. This method has the advantage that it does not necessitate the user entering every defining variable each time a component is drawn.

24. Components are created and held in the drawing file using a 'tree' data structure, as illustrated in figure 8. The advantage is that the elements forming a component can be referenced in a simple manner. For example, if a component is to be shown on a general arrangement drawing the branches 'Dimension & Grid' and 'Reinforcement' will be omitted when loading the component.

25. When a component is drawn its defining variables are written to the drawing database as part of the tree structure. This is shown in figure 8. as the branch 'Component Definition and MTO'. As well as providing data for material take-off, this information provides a means by which the user can query a component and reset the defining variables to create automatically an alternative view. Thus a section or second elevation view can be created at any time, without any data re-entry simply by identifying the component on the screen.

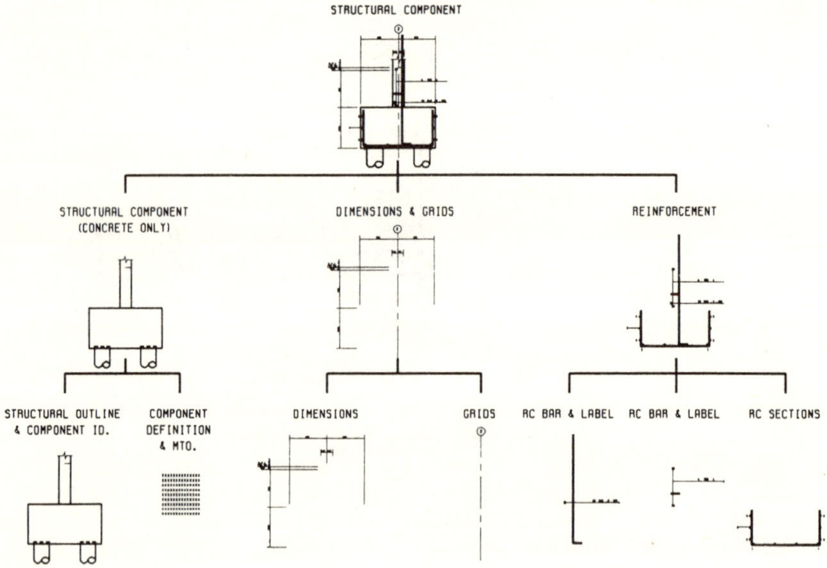

Fig. 8. Typical component data structure

26. Component integrity (i.e. correctness and completeness of the data structure) is maintained by the use of protection 'switches' preventing the user from accidentally deleting or modifying components in an 'illegal' manner.

FUTURE DEVELOPMENTS

27. In addition to a continued expansion of the libraries of standard details there are 3 main areas where development can be foreseen:-

(a) The storage of component definition data. Each component has a considerable quantity of associated non-graphical data for component integrity and material take-off. This adds to the size of the drawing sheet file and degrades workstation response time. It is intended that this information will be stored in an 'off sheet' database, and methods for achieving this are under investigation.

(b) The use of 3-D modelling. In the majority of circumstances (especially in building frames) a 2-D approach to structural r.c. detailing is more than adequate. It has the merits of economy and responsiveness. However there are undoubted benefits to be gained from 3-D modelling of complicated, heavily reinforced joint details. Work is under way to enable 2D details to be used as automatic generators for 3D joint models.

(c) Links to design programs. The modular component approach which has been adopted lends itself to linking to specialist programs for r.c. design. This could involve in-house developed software or suitable commercially available packages.

CONCLUSIONS

28. The introduction of CAD into the Structural Design Department for r.c. detailing has been successful. Considerable effort has, however, been required Full discussion took place with the Department during the specification and implementation of the RCCD package, and there is now full departmental awareness of the capabilities of CAD and acceptance of the new technology.

29. CAD has to coexist with manual detailing, a situation which will continue for some years yet. The drawing and schedules produced manually and by CAD must, therefore, have similar appearance and meaning.

30. The Structural Design Department is benefitting from this r.c. customisation. The overall productivity is estimated to be four times faster than manual draughting with some tasks even being an order of magnitude faster.

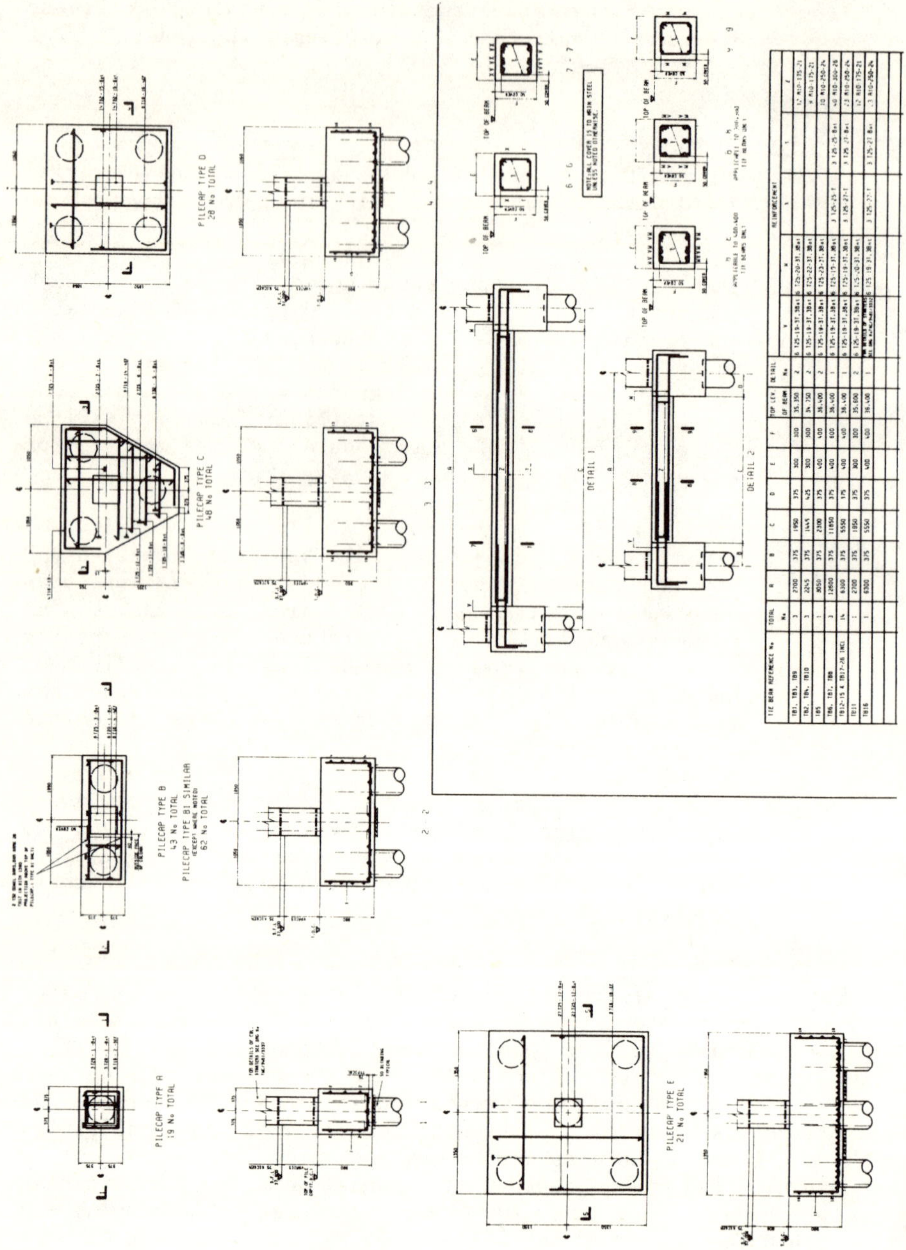

Fig. 9. Typical Drawing using RCCD Package

Paper 5. The establishment and development of CAD in the public sector

P. JONES, MICE, DipTE, G. HOPPER, MIStructE, and
J.G. BERRESFORD, BSc, Transportation and Engineering Department,
Dorset County Council

SYNOPSIS. This paper deals with the development of a CAD system in the Transportation and Engineering Department of Dorset County Council. It includes a discussion on how a supplier was selected, the training involved and future developments which are being considered.

THE DEVELOPMENT OF A CAD SYSTEM

1. The Design and Construction section of the Transportation and Engineering Department undertakes the planning and design of civil and structural engineering works and supervises the construction of these works. The workload comprises:-

(a) Highway planning, design and construction; design of highway structures including road and footbridges, retaining walls and pedestrian subways.

(b) Structural design of buildings and their components in reinforced concrete, steel, timber and load bearing brickwork.

2. As a Local Authority, Dorset County Council is dealing with increasing demands on all services including the design office. The County is an area of high expansion leading to more demands for development and the associated infrastructure. These demands require a proper co-ordinated professional response and a continuing improvement in the quality of the service, whereby it has become more necessary to accelerate production rates of information, and in particular that related to many of the repetitive elements of the planning, design and construction processes.

3. The Planning and Transportation Committee at their meeting in January 1983 accepted that provision should be included within the future capital budgets for the provision of a Computer Aided Design System (CAD system). It was envisaged that the introduction of the CAD system would:-

(a) Increase efficiency and competitiveness
permit schemes to be designed three-dimensionally
permit detailed evaluation of all designs and options
reduce leading in time for capital projects
help to cope with the proliferation of Codes of Practice and Regulations
assist with the increasing complexity of modern design work
deal with demands for detailed information
help to resist claims from contractors who are increasingly insisting on information being supplied on the basis of a computer format
improve quality of design and drawings
provide for improved service both to the County Council and to Government departments
reduce cost of service

(b) assist to meet the overall capital budget programme which was continuing to expand, and the post-1985/86 figure was expected to be at least £15m.

Table 1. Capital Expenditure

YEAR	1982/83	1983/84	1984/85	1985/86	1986/87	1987/88
GROSS CAPITAL	£11.4m	£14.3m	£14.6m	£18.9m	£19.3m	£19.6m (Estimated)

(c) assist with the overall resources/staff situation, enabling the retention of staffing resources at about the existing levels, provided that CAD could be introduced within the 1984/85 period, whereby no action would be taken to fill a further ten Design and Construction posts.

THE APPRAISAL OF CAD SYSTEMS

4. Prior to obtaining the Planning and Transportation Committee's authority to proceed with the implementation of the CAD system, a group of engineers and technicians from the design and construction section had appraised a number of products available on the market. This was coupled with discussions with several established CAD users and an overall view of what would be expected from the CAD system together with short-comings was essential in order that a reasonable specification could be achieved.

5. In the course of the investigations a few facts rapidly became clear, not the least of which was the relatively cautious investment in CAD by the UK construction industry compared with the USA and Europe, or indeed with other industries in the UK. Clearly, in mechanical and electrical engineering, development costs of one item tend to be recovered over a large production run, whereas this is not generally true of civil engineering.

6. It was also apparent that, if one merely wished to replace the drawing board with its electronic equivalent, then virtually every system would be suitable, as all that would be required were a few geometric constructions and the facility for writing text on the drawing. It was equally apparent, however, that far more was on offer from CAD systems that could be of real practical value to the engineer, and also to the project administrator.

7. Of particular interest was three-dimensional modelling. Many is the time that the engineer has had to resort to rudimentary cardboard or polystyrene models of complicated details, in order to ensure that what he shows on the drawing is a true representation of the structure he wishes to build. Given 3-D modelling, the representation of a structure in the form of 2-D plans, elevations and sections ought to be far less error-prone. Furthermore, the representation of perspective views of a scheme would be most useful at the public participation and committee stages, where one is usually dealing with people who are not used to interpreting the finished product from a set of drawings. However, it was found that very little use was made of 3-D modelling in civil engineering.

8. Care was taken not to write too rigid a specification which would have precluded all the available systems, and the software specification required the following:-

(a) The MOSS highway design suite of programs which was seen to offer a considerable saving in drafting time, as it takes away the labour-intensive task of the production of longitudinal and cross-sections, so the ability to produce MOSS drawings received a high priority in the specification for our ideal system.

(b) The advantages of "layering" a drawing, so that the graphics for example on a concrete outline drawing can later be used for the reinforcement details, were also recognised and included in the specification.

(c) Whereas repetition is not very common in civil engineering, it does occur with certain items such as bridge beams and parapet posts, therefore an object library was considered an important feature.

(d) Reinforced concrete detailing with automatic bar bending schedule was required in the system. Investigations had shown that, although a number were available, no one system could really be said to do everything. Accordingly, a benchmark test was set which would prove a system's ability to cope with the majority of schemes.

(e) As far as general drafting was concerned, a suitable performance test was found which, it is understood was originally developed in the Netherlands. This test, referred to as 'the test for electronic draughting is described in Appendix 1 and was found most useful in assessing system capabilities.

9. The decision was therefore taken to proceed with a CAD system which should provide for:-

(a) general drafting
(b) the MOSS system
(c) reinforced concrete drawings with automatic bar bending schedule production
(d) three dimensional drawing
(e) colour.

THE ACQUISITION OF THE CAD SYSTEM

10. Some £200,000 had been budgeted for the acquisition of a CAD system which was anticipated to allow for two to three work stations. Whilst various CAD systems had been appraised it was necessary in accordance with the County Council's Standing Orders to place public advertisements in the technical press and "The Times", inviting manufacturers/suppliers to apply for the CAD System Select Tender List. Overall, 22 applications were received and 9 companies were eventually placed on the tender list and tenders were to be returned by 1 March 1984.

11. Following the receipt of tenders, a further series of tests were undertaken to assess the performance of the equipment offered and three systems were subjected to a formal "test for electronic drafting". Details are described in Appendix 1. On the completion of these tests, it became apparent that no tenderer could fully meet the specification and it was decided to invite new tenders for the CAD system in August/September 1984. During the interim period further development work was being undertaken by the potential suppliers.

12. On receipt of the new tenders from the nine companies further tests were undertaken and a tender from Applied Research of Cambridge Limited (ARC) was accepted in April 1985, totalling £192,260 which included licences and initial maintenance and provided for:-

(a) <u>Hardware</u>

Central processing unit (5MB)	VDU
3 numbers 71 MB disk units	Plotter
printer	Screen copier
Communications interface	3 number work stations

(b) <u>Software</u>
Software for the drafting system
Three dimensional viewing
Reinforced concrete detailing system

(c) Training for nine members of staff was inclusive and rates for training additional personnel were set out in the tender.

13. The equipment was installed in August/September 1985, together with the MOSS system, which involved a further £16,127.51 including the first year's maintenance. Two further work stations, including a colour station, were added in June 1986 at an overall cost of £36,365.00. Two further similar work stations are currently being added and in the next financial year a further two to three work stations should be added together with an additional networked central processing unit. These additions require tenders to be invited in the normal way in accordance with Standing Orders. Figure 1 below shows the system configuration.

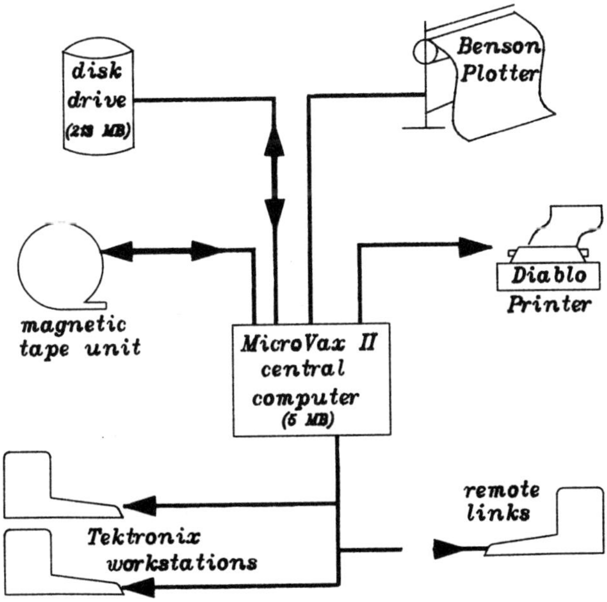

Fig. 1 System Configuration

14. ARC became part of the McDonnell Douglas group in the last quarter of 1985. The changeover has led to the provision of an improved standard of software development, though service is more impersonal with the larger company.

TRAINING

15. The training schemes offered by ARC/McDonnell Douglas were fully utilized and two parties of eight trainee were sent on courses over a period of nine months. Neither of the groups had any working experience of CAD but the second group benefited greatly from the experience already obtained by the first and this was shown by the speed with which the later group of trainees developed. Experience therefore indicates that the

time lag between hardware installation and productive drawing output could have been much reduced had an experienced CAD operator been available from the start. Despite this and because staff at all levels gave the system an enthusiastic reception production of working drawings was in advance of original expectations.

16. Training has been undertaken at Cambridge and Woking which has involved the staff attending courses of up to five days duration, and the training for MOSS has been undertaken in-house. Further training will continue to be provided and to date the following staff training programme has been completed:-

(a) GDSA Course
8 staff Sept 1985
8 staff Mar 1986

(b) GDSB Course
8 staff Nov 1985
4 staff May 1986
2 staff Aug 1986

(c) System Managers' Course
2 staff Oct 1985

(d) 3 "D"
4 staff Nov 1985

(e) Reinforced Concrete Detailing
4 staff Feb 1986

(f) MOSS
20 staff Oct 1985

17. The staff undergoing the training courses have mainly been technicians and assistant engineers. For the MOSS course team leaders were included whilst the systems managers are senior assistant engineers.

DRAWING PRODUCTION WITH THE CAD SYSTEM

18. Within seven months of the training period two major road and bridge schemes totalling £11.0m were put out to tender and 60% of the drawings were produced with the CAD system. Prior to inviting tenders a critical situation arose due to the need to carry out final amendments to drawings and at the same time produce plots of the tender drawings and to record all finished tender drawings in the archives. This situation was overcome by adopting an overtime plus shift working system and the formalisation of a better planned approach whereby priorities can be identified at peak demand times on the CAD system. In particular it is essential to ensure that the drawings in the archives are accurate and register up-to-date amendments.

19. The highway drawings are generally created by the MOSS system in conjunction with digital aerial survey data and these are transferred to the General Drafting System and filed for future recall. With the MOSS system, in addition to the drawing creations, the taking off of quantities is also undertaken together with the provision of other important information such as setting out data in the form of co-ordinates or dimensions.

20. Structural drawings usually divide into two main areas,

bridgeworks and associated highway structures, and building structures. The Department acts as a Consulting Engineer to the County Architect in structural engineering matters. A large amount of structural drawing in this respect has been produced on the CAD system. For example a technician and say an engineer can be working simultaneously on different parts of the same drawing, thereby saving time without any loss of drawing style or accuracy.

21. There was an initial tendency for the bridgeworks general arrangement drawings to take longer to prepare on the CAD system as compared with the normal drawing board approach. However, once the general arrangement drawings have been properly constructed the associated drawings such as the reinforcement detailing through the reinforced concrete detailing system are produced much quicker and with greater accuracy. The opportunity for the designer to take hard copies of the drawings at all stages of development means that drawings can be continually monitored whilst being produced.

FUTURE DEVELOPMENT OF THE CAD SYSTEM

22. The Department is continuing to develop a library of objects which are available from screen menus thereby reducing the time required to produce drawings. A further advancement involves the development of input files which generate about 85% of the drafting content, say for example a pedestrian subway taking overall less than a working day to produce.

23. The GDS system includes GDS BASIC, a programming language which permits a limited amount of in-house development, as well as being useful for the creation of parametric components. This has now been supplemented by EPI (External Program Interface), which enables external Pascal or Fotram programs to be interfaced to the basic GDS system, opening up a wide field of development whereby further automation and better management of the processes involved in planning and design is being achieved.

24. McDonnell Douglas Information Systems Ltd continue to show a very great commitment to the civil engineering industry including an intensive development programme. The company met the specification and their services and back up support has not let the design and construction section down. At the same time area group meetings are now beginning to show results regarding the improvement and further development of the CAD system and the work being done corporately in Dorset will benefit all those concerned with the design and construction industries.

SUMMARY

25. The overall case for the CAD system discussed in paragraph 3 has been met but there has not been the expected progress regarding the production of Codes of Practices and Regulations being provided on tape or disc format. The design and

construction section continues successfully to compete for project design work.

26. The pre-investigatory work undertaken by all levels of staff who were expected to and are now using the CAD system was very worthwhile. Having bought what we wanted and the staff remain committed to it.

27. The total cost of the CAD system is likely to be £400,000 which is about 2% of current annual gross capital budgets and if the cost is spread say over five years it equates to less than ½% per annum. On top of this there is the maintenance costs equivalent of up to four staff posts. Therefore the final word must of course rest with the three "E"s, and on the basis of our present experience it is Effective, Efficient and has made Economies because it:

(a) avoids staff doing repetitive work, improves accurancy and reduces mistakes

(b) its production rates continue to improve without any loss of quality

(c) it has saved in the order of £200,000 per annum on staff costs.

28. We can only suggest that our approach in Dorset to make full and proper use of modern technology is helping to provide all our customers with a better service and at the same time is enhancing the standing of the civil engineer in the highly competitive professional fields within which we operate.

Acknowledgement is made to Mr G L Vizard, County Surveyor, BSc DipTP(Lond) CEng MICE MRTPI ASVA MInstWM FRSA FIHT DipTE.

APPENDIX 1

COMPUTER AIDED DESIGN - TESTS

1. The 'test for electronic drafting' required the drawing shown in Figure 2 to be produced in various layers. Drawing with the aid of a grid was prohibited. Overall the test was required to show:-

(a) various line styles and objects
(b) that parts of the drawing could be interchanged and amended between layers
(c) that parts of the drawing could be replotted, transferred, rotated and re-scaled
(d) areas and dimensions of the drawing could be generated automatically.

2. The reinforced concrete detailing system required the production of a drawing for a reinforced concrete wall fully detailed with a metric bar schedule being produced automatically including shape code 99.

3. In order to assess the response time and memory usage, various types of lines and lettering were required to be drawn with an indication of the necessary disc storage space together with the time taken to produce the drawings onto the screen.

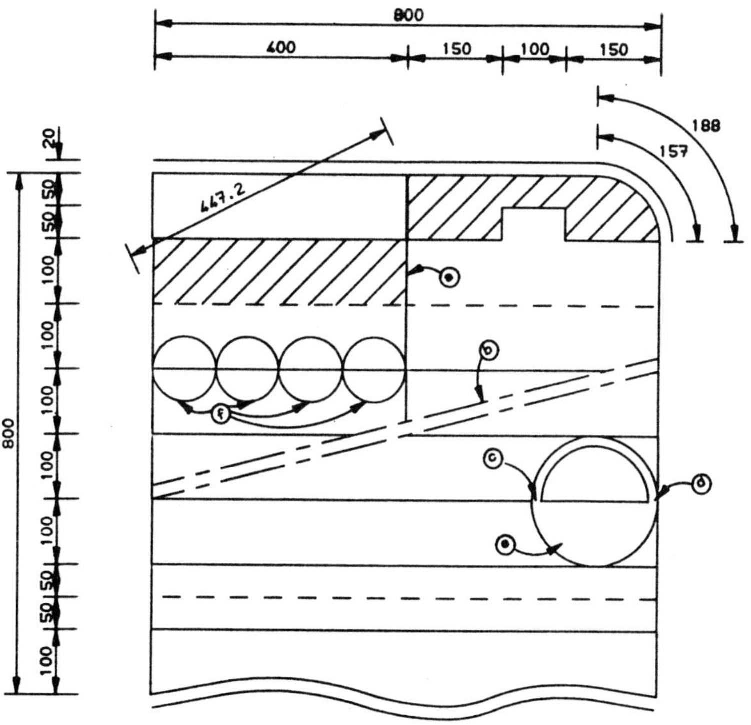

Scale 1:10

Paper 6. Reinforced concrete detailing systems for civil engineering

J.R. ROWLEY, BSc(Eng), MICE, and J.C. NEWBY, BSc(Eng), MICE, MIStructE, W.S. Atkins & Partners

SYNOPSIS
WS Atkins & Partners is a large civil engineering consultancy responsible for the design and detailing of a wide range of reinforced concrete structures. Over the past few years, the consultancy has introduced computer aided draughting and in the process has investigated a number of alternative systems capable of RC detailing. In-house trials of three systems and demonstration of others have enabled the authors to produce a list of the functional requirements that make an RC detailing system feasible for use in Civil Engineering.

INTRODUCTION

1.1 The authors represent the CAD department of WS Atkins & Partners which provides CAD support to divisions covering civil, structural, transportation and water & public health engineering. Each of these divisions and disciplines produce reinforced concrete designs which have their own particular requirements for differing functions within an RC detailing system. Against this varied background, three systems - BARD, GIPSYS and GDS have been used within WS Atkins & Partners for RC detailing. This paper does not seek to compare these nor any other systems, instead, the experience gained is used to discuss the requirements of a reinforced concrete detailing system for civil engineering.

1.2 Several suppliers were found to offer detailing packages linked to design programs. Examination of these showed that in general they applied only to simple regular concrete elements; beams, columns, slabs etc, and that the design aspect covered building codes only - at that time CP110. Projects undertaken by WS Atkins Structural Division fell into this category and BARD, produced by the C&CA was purchased and has since been used successfully on several projects.

1.3 Projects undertaken by other divisions did not fall into the BARD category due to irregular elements and the use of non-building codes. A general RC detailing system was therefore required that would handle all these other projects. WS Atkins chose to investigate GDS and GIPSYS by extensive trial - GDS because it had been in use in the firm for several years as a general draughting system and GIPSYS because, specified and written by Consulting Engineers, it seemed likely to be the more comprehensive solution. The comments that follow relate to RC detailing systems for general use.

FUNDAMENTAL REQUIREMENTS

2.1 The objectives upon which the selection of a system was based were fundamentally threefold; improved quality, consistent presentation and cost savings.

2.2 Quality is improved beyond that achieved manually by many CAD features; e.g. automatic scheduling and replication of details. The automation of as many error prone processes as possible increases reliability; e.g. the use and update of outlines shown on G.A.s eliminates possible error in redrawing for RC detailing.

2.3 Improved presentation is obtained through uniformity and neatness but does require that the whole of the drawing is produced on the CAD system. The ability to easily compose RC drawing outlines at different scales and content to the equivalent G.A.s also works to improve the presentation of congested details.

2.4 Cost savings are a function of the capital cost of the system, the investment in training, management charges, the reduced time taken to produce a drawing and the amount of use made of the system.

SOFTWARE FUNCTIONAL REQUIREMENTS

2D, 2½D or 3D

3.1 Detailed discussion of the differences between the models created within CAD systems is outside the scope of this paper. Nevertheless, RC detailing system suppliers have adopted a variety of approaches. The simplest is to take the view that the system replaces a drawing board and then build on that concept so that features such as automatic scheduling are included. A particular drawback of this approach is to limit views of bars to those on orthogonal axes.

3.2 The other extreme is to adopt a 3D model approach wherein the structure to be reinforced is represented either as elemental solids or as a wire frame. The reinforcement is then input to the system relative (usually) to the surfaces of the object. The 3D model in general civil engineering applications is costly to generate but it can be quite useful to be able to work on a single isometric view of the object when positioning 3D bars. The main problem with the approach, however, is that the 3D model generated is not a conventional description for civil engineering drawings. The automatic generation of conventional plan, elevation and section views may be available in some systems, e.g. Laing Technology Group's Computervision, but the operator is still faced with the choice between making further changes on the model or taking a shortcut by altering only the generated views. The authors' view is that, in many cases, 3D is likely to provide as many problems as it solves.

3.3 Some systems fall between the 2D and 3D categories and as such are often referred to as 2½D systems. GIPSYS, for example, includes the ability to move bars within a 3D axis system and place then onto a 2D outline drawing. PAFEC's DOGS RC has the ability to produce automatic cross-sections from 2D definitions of simple elements.

Codes of Practice

4.1 Most detailing systems incorporate support for BS4466. This includes a definition of the preferred shapes but quite possibly not all shapes. In simpler systems, none of the 3D shapes are likely to be available. The authors have found the GIPSYS approach, using an external file to list the practice dependent algorithms, to be a useful one. Computing and advanced geometric skills are necessary to be able to generate or add to this external file but the advantage of being able to support any project requirement makes this worthwhile. The file also allows user defined checks for bending dimension compliance to be included.

Non-Standard bars

5.1 Whilst support for standard BS4466 shapes is always included, the way in which systems deal with non-standard shapes varies considerably. Two types of non-standard bars are required; those used regularly and those produced for one-off occurrences.

5.2 WS Atkins use some 12 non-standard shapes for bridgeworks and thus require these to be available parametrically and to appear on a schedule as shape code 99103 for example. With GIPSYS this has been achieved by adding to the external code of practice file. Sample bars are shown in figure 1.

5.3 When one-off non-standard shapes are required they should be created graphically within the detailing system rather than forcing an exit to edit the external file.

SHAPE CODE	METHOD OF MEASUREMENT OF BENDING DIMENSIONS	TOTAL LENGTH OF BAR (L) MEASURED ALONG CENTRE LINE	DIMENSIONS TO BE STATED IN COLUMNS A to E OF SCHEDULE
99102	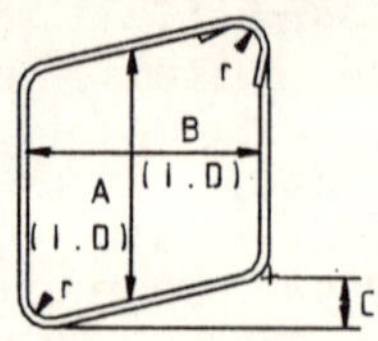	2(A+B)+3n	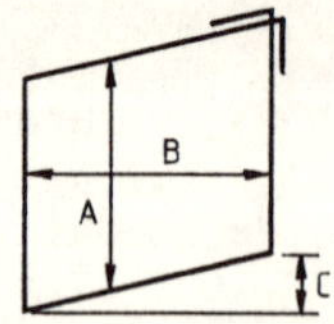
99105	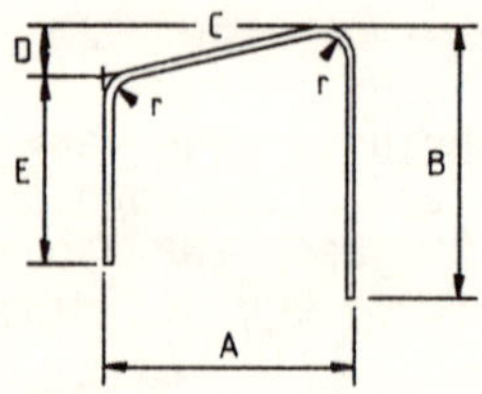	B+C+E-r-2d	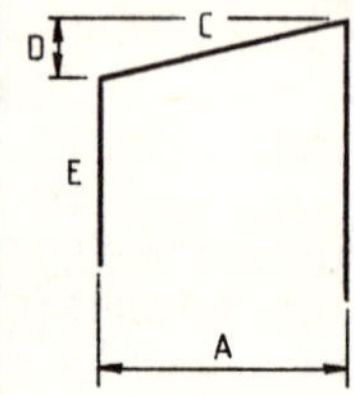

Figure 1: Sample non-standard bar shapes

<u>Specifying bar dimensions and angles</u>

6.1 The most regularly used method of specifying bar dimensions is to key them in. Otherwise, dimensions, from which cover is deducted automatically, and angles are indicated by hitting the outline. This process yields a version of the current bar which is fully dimensioned but not yet placed. Most systems seem to use this approach of pre-defining the bar dimensions before placing.

<u>Bar placing</u>

7.1 The placing of bars on a drawing is usually carried out by reference to one of the corner origins. The user should be able to choose any vertex as the origin and, in addition, should be able to snap the mid-point along any leg onto any feature line of the drawing. Other placing requirements are to mirror or rotate the bar about any axis.

<u>Covers</u>

8.1 Bending dimensions can be derived reliably by the use of pre-specified covers. The GIPSYS external practice file allows covers to be given variable names e.g. FC - foundation cover, and then used in the input of bar dimensions. Corner origins may also be offset by cover dimensions in order that bars may be placed on an exact point such as the corner of an element.

8.2 GDS employs cover boundaries in which the cover outline is traced inside the element outline. Bar dimensions may then be fixed by this boundary foregoing the need to subtract covers from overall dimensions. Covers may be checked whenever necessary by the 'measure' facility but the authors prefer to have the information pre-defined as with GIPSYS.

Lap lengths

9.1 The ability to store and use standard lap lengths is another aid to improved reliability but as with bar shapes these need to be adjustable to suit the particular code that is being worked to. Again this should be a function of the external practice file so that consistent usage is assured.

Ladders

10.1 Most call-up extents are indicated by a simple 'ladder' showing one view of a sample bar which is intersected orthogonally by the ladder line. Conventially, the call-up is placed at one end of the ladder. However, if the ladder arrangements are limited to this basic concept it would be impossible to prepare complete reinforced concrete drawings.

10.2 The requirements for more complex ladders may be divided into purely graphical changes and those which affect the placing of bars. Examples of the former would be joggling the end of the ladder to find room for otherwise cramped call-up text, or placing bars at an angle other than orthogonal to the ladder. The other type includes variable (a-z) bars, sub-ladder lengths with different spacing, radial and circumferential ladders. An example of the use of variable bars is shown in figure 2. It helps keep the system simple if the elementary, i.e. graphical, changes can be obtained with standard detailing rather than special RC detailing commands.

Call-ups

11.1 In the authors' experience, little deviation is now made in the UK from the usual call-up notation of; number, steel grade, diameter, bar mark, centres and position. e.g. 15T25-10-200B. However the ability to user-define call up style may be necessary for overseas work.

11.2 The ability to add additional information to call-ups, such as NF, FF, ABR etc is required. Annotation of further drawings of bars in section is also necessary and can take one of several forms as shown in figure 3. Measured dimensions necessary for steelfixing should be obtainable from the graphical database.

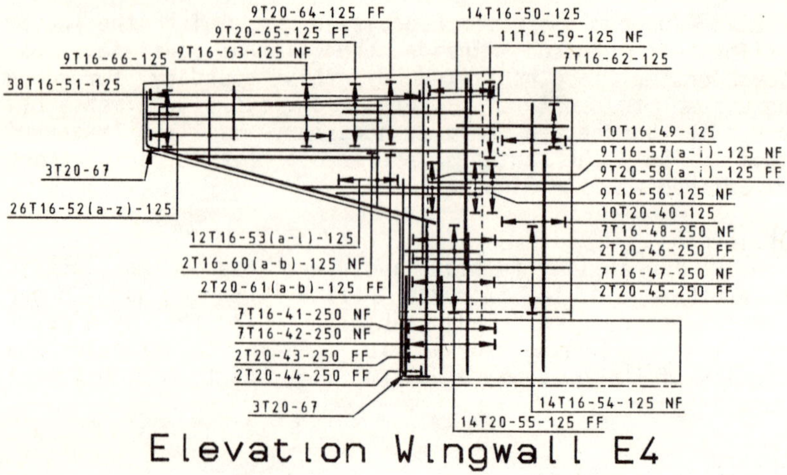

Figure 2: Example of the use of variable bars

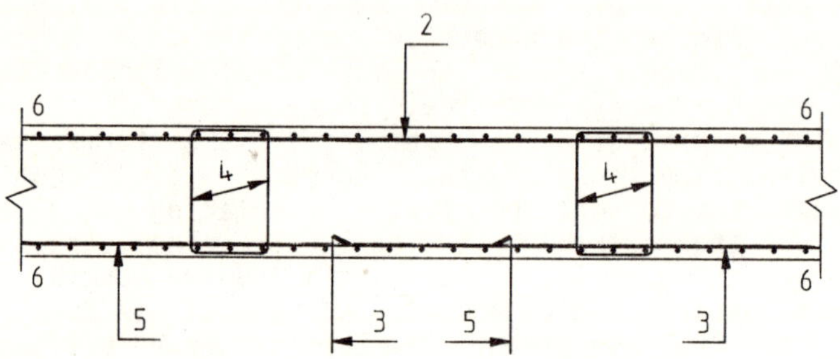

Figure 3: Bar Call-ups - secondary annotation

Draughtsman's licence

12.1 With general arrangement draughting there is only one database in use - the graphical one. Thus, draughtsman's licence is accomodated by simply drawing as one wishes to see. RC detailing requires an additional database for storing the scheduling data and it is important that this is not affected by visual improvements. Examples would include a detailer reducing the drawn length of a bar to simplify a complex detail. This is handled quite easily in a system where the bars are held in 3D as the bar may be rotated into the paper to reduce the viewed length but holding the true length unchanged. Further drawings of the bar based on database, rather than graphical information, will result in the true length being retrieved. 2D systems such as GDS allow the user to trim leg lengths with graphical editing commands which do not affect the scheduled length but can affect subsequent copies.

12.2 Other examples include the choice between showing 'ticks' at either bar end, or not, if so required and the diameter of 'blobs' representing bars in section which, irrespective of scale, should not fall below a minimum of approximately 1mm.

12.3 The ability to amend graphics whilst leaving the schedule database unchanged is required for the not uncommon practice of scheduling a straight bar but showing it fixed to a curve on the drawing.

Ease of editing

13.1 The ease with which changes can be made to the detailing is important for productivity but editing facilities must also ensure that the inherent reliability of the system is maintained. Deletion facilities, for example, must delete all drawings of a bar not just the one in the view being worked on. Changing of bar diameter must adjust radii to suit. Effects of such a change of diameter on lap length are usually left to the detailer to adjust although it is conceivable that, given an appropriate database structure, this could be carried out automatically.

13.2 Editing facilities often appear to have been added as an afterthought rather than as an integral part of an RC detailing system as anything other than minor changes result in bars having to be completely deleted and then input again from scratch. As a result it can take longer to make a fairly minor change to existing bars than it took to input them in the first place.

Bar group copying

14.1 In practice, bars are grouped together for scheduling purposes and this facility is included in most systems. The reuse of groups of bars by copying, mirroring and rotating requires the option to either maintain bar mark numbers and add to existing total numbers of bars or to change the bars to a new set of bar marks. In the first instance changes (including deletions) to a bar in one group should be reflected in the copied group but in the second instance, total independence, one group from the other, is required. It is, of course, obvious that full 'tracking' of mark numbers to avoid duplicates is essential.

Schedule format and contents

15.1 Schedules are generally formatted in accordance with BS4466 although the facility to amend to suit house styles or project requirements is necessary.

15.2 The summation of bar length by diameter and/or other category and thus calculation of weights of reinforcement is a simple computer operation and necessary for Bills of Quantity. These summations are generally carried out for each drawing, which is quite adequate for most circumstances. However, the Department of Transport Method of Measurement for Road and Bridgeworks (Section 15) requires that reinforcement weights are summarised by structural element category, e.g. end-supports sub-structure, intermediate-supports substructures, super-structures etc. These categories are often not identifiable by a drawing split, rendering the drawing weight summaries redundant. Providing the groups are named in accordance with the Department method it should be feasible to schedule, and therefore weight, a whole group of drawings as a batch.

Hardware

16.1 Most of the available RC packages run on mini-computer or workstation hardware platforms. Capital costs, including software are still in the order of £25,000 per 'seat' although competition and technical development is having a remarkable effect on the hardware component. At the other end of the scale, the less functional RC packages associated with micro-computer based CAD stations are now becoming available on micros which are potentially more powerful than typical minis. Here the capital costs can be as low as £10,000 per seat.

16.2 In the authors' experience, a well thought out use of colour can greatly enhance an RC detailing system's ease of use.

CONCLUSIONS

17.1 RC detailing is not as simple an exercise as at first it might appear. The authors' impression is that all software houses initially underestimated requirements and that while some are now putting right many of the initial deficiences, others have still to act.

17.2 RC detailing of civil engineering structures requires much flexibility within CAD systems. The facility for the user to tailor a system to his own requirements is important, indeed vital, for many businesses.

17.3 No system as yet offers the definitive single solution for all users. Nevertheless, speaking from experience, it has been possible to complete all drawings using the CAD systems available.

Paper 7. Computer aided detailing — the current context

A.S. WATSON, BTech, PhD, MICE, Department of Civil Engineering, University of Leeds

SYNOPSIS. This paper is not concerned with specific detailing systems; it seeks to establish the context in which such systems operated by exploring the wider field of Computer Aided Design. Most computer detailing systems have developed in parallel with design and/or drafting sytems. The reasons for such linkage are considered and conclusions drawn regarding the future evolution of detailing systems.

INTRODUCTION

1. Many civil engineers have a sceptical view of the technical application of computers. This attitude can be defended since the majority of structures lack features that make computer techniques essential for the production of a good design.

2. If unusual structures and specialist sectors of the industry are excluded, computers are applied to the design of structures only when it is believed to be cost effective. However such evaluations are not easily made and decisions are also influenced by considerations of image and personal preference.

3. Much pioneering work on the application of computers to the process of structural design was conducted by civil engineers. While technical limitations remain, the continuing fall in the real cost of computers and the constant pressure on fees will progressively increase the proportion of work conducted with the aid of computers.

THE PROCESS OF STRUCTURAL DESIGN

4. Most structural designs break down naturally into the three phases; preliminary design, analysis and detailed design. Interactions with other members of the design team may limit the sequential nature of the process (Ref. 1).

Preliminary design

5. The most critical design decisions are made during the preliminary design; the structural form, materials and probable member sizes. Computers have made little impact on this phase since it is difficult to computerise, drawing as it does on the experience and intuition of the engineer. Also it is the

most creative phase of the design process and thus the one in which engineers would least like to see the intrusion of computers.

6. However there is agreement that decision support systems applied to this phase of the design process offer the greatest potential for improving the quality of design. Practical computer tools will not become generally available for some time and are likely to be based on Artificial Intelligence techniques (Ref. 2,3).

Analysis

7. Having completed the preliminary design the process of structural analysis is relatively deterministic and thus more amenable to solution by computer. While changes in design codes have tended to increase the volume of analysis the majority of structures can still be analysed by hand. However there is general agreement on the value of computers for the analysis of complex structures.

8. Significant improvements in the user interface of the current generation of analysis programs have reduced the real cost of their use to the point that it is now a cost effective option for the majority of structures.

Detailed Design

9. Detailed design covers the design in detail of each structural element, the preparation of working drawings and the preparation of specifications. This phase is critical to the correctness of the design yet it is a major production task. Traditionally considerable reliance was placed on the skills of technicians to assist in the execution of this work. The current lack of experienced technicians is a significant factor in the growing interest in the use of the computer as a production tool. However the application of computers to detailed design phase is not as yet sufficiently widespread to indicate general acceptance of the benefits.

10. Within specific fields specialist software packages have become established design tools, for example MOSS (Ref. 4) for highways and major earthworks applications. This is not the case for the general design of reinforced concrete and steelwork structures where the use of computers as a production tool is variable but overall is still at a low level. For example, most specifications are now prepared with the aid of a word processor but the proportion of drawings prepared using a computer drafting system is small.

11. As the economic factors become less significant it is becoming apparent that other issues are delaying the general deployment of computers as a production tool.

DETAILING

12. Detailing is the process of generating the low level information necessary to allow the construction of individual structural elements. It takes place after the general arrangement of the structure has been defined and the overall

structural analysis completed and follows on from, but may be an integral part of, the detailed design of the elements. Usually the information is expressed in the form of drawings and schedules. Their preparation requires considerable skill and experience and frequently involves the need for supplementary calculations. Significant differences exist in the nature of detailing as applied to reinforced concrete and structural steelwork.

Reinforced Concrete

13. UK designer offices traditionally produce to-scale reinforcement detail drawings plus bar schedules. These may be prepared by the design engineer or by a detailer (who may be located elsewhere). Thus the design calculations for each element tend to be comprehensive and accurate general arrangement drawings are a pre-requisite.

14. In some cases the reinforcement is detailed by the contractor, the designer usually provides typical details and calculations that specify the amount and position of the reinforcement.

15. The detailer must ensure that the reinforcement satisfies the designers calculations, is well detailed and is practical to construct. To do this he needs access to the calculations and the general arrangement drawings but there is little need for interaction with the designer. Greater interaction may be needed if the reinforcement is unusual or congested.

Structural Steelwork

16. Usually structural steelwork is detailed by the fabricator, the design engineer supplies schematic drawings which indicate member sizes and design loads for the connections. The design of the connectors is the responsibility of the fabricator although the designer will probably indicate the form of the connection. In some situations the design engineer will prepare fully detailed drawings.

17. The steelwork detailer produces fabrication drawings and calculations for the connections. He is concerned with fabrication and erection as well as the structural adequacy of the connections.

Summary

18. The detailing of reinforced concrete can be sub-divided into two relatively independent operations. The preparation of detailed design calculations, which interacts quite strongly with the other design phases, and the production of detail drawings which does not. Steelwork detailing normally does not interact strongly with the other design phases but the production of detailed design calculations and fabrication drawings are usually closely linked.

COMPUTER AIDED DESIGN SYSTEMS

19. Currently most software packages can be classified as

analysis, elemental design or drafting. Analysis programs have evolved over many years and are now very sophisticated but as yet they do not offer guidance to the user (Ref. 5).

Elemental Design

20. Programs for the detailed design of indivdiual structural elements have been available for some time but have not achieved widespread application. A major factor is their lack of flexibility. To maximise the efficiency of use the volume of input data is usually kept to a minimum which restricts the potential scope of the program. Thus the software may only be able to deal with a proportion of the members in a structure. This disrupts the production cycle. Also, while the resulting design may be adequate, it may be inconsistent in detail with other parts of the structure.

21. Automation of the design process raises philosophical problems since they are many correct solutions. Thus, if significant savings are to be made by the application of computers, this may necessitate some reduction of the engineer's control of the solution. The correct balance may not have been achieved by the current generation of design packages but a degree of delegation is inevitable. It can be inferred that confidence and communication are important issues.

Computer Drafting

22. Some UK companies have invested heavily in computer drafting systems but such systems have yet to make a significant impact on the production of structural drawings. It is inevitable that this situation will change. Cost has been a major obstacle and, while this is no longer the case, the UK market for construction industry drafting systems is both small and fragmented (Ref. 6). One consequence is that vendors have not had the revenues for rapid product development and many of the systems currently available were not created for construction industry applications. Many drafting system users are not able to realise all the potential benefits of computer drafting (Ref. 7).

23. Drafting Systems can be classified into a number of generic types - 2D, 2½D, 3D and Geometric Solid Modelling (GSM) - depending upon the nature of the database used to store the linework. Both the capability and the cost of systems increase in moving up this range from 2D to GSM.

24. The most significant of the potential benefits is concerned with coordination, some systems enable the creation of compliant drawings which have linework that is always mutually compatible. 2D systems are generally the simplest to use but in this case the potential for automatic compliance is limited to drawings lying in the same plane. Little compatibility exists between the databases of different drafting systems, even if they are of the same generic type. The Initial Graphics Exchange Standard (IGES) (Ref. 8) offers only a partial solution to the problem of transferring data between different systems. Thus computer drafting systems do

not, as yet, provide a general means of improving inter-office communications.

Integration

25. Free-standing systems for the design or detailing of structural elements have advantages but their weakness lies in the volume of data needed to fully define the problem. Most systems avoid the problem by only dealing with a narrow range of element types and/or by restricting the scope of what is performed. For example, by not producing to-scale reinforcement drawings.

26. The integration of systems to minimise the need for man/machine communication is the solution. A notable early application of this philosophy was the Genesys (Ref. 9) sub-system RC-BUILDINGS. Conceived in the early 70's before interactive computing was common, this package took a "black-box" approach. An advanced problem orientated command language was used to define the geometry, member sizes and loading on the frame and the package would then proceed to analysis, design and detail the structure. Details where based on standard arrangements, only bar bending and fixing schedules being output. Later versions also plotted general arrangement drawings. The system worked but a significant amount of manual work was needed to deal with the areas with which RC-BUILDINGS was unable to cope.

27. Interactive computing facilities have enabled more flexible systems to be created in which integration is used to avoid the need for users to re-input data. One technique is to loose couple two or more software modules so that data from one is passed onto another. In this way an analysis module may pass its input data and results on to a design module and this in turn can pass its data on to a detailing module. The user retains interactive control over each module and the flexibility to still use them individually. This is a pragmatic approach to integration but the ad-hoc basis restricts the scope for long term development.

28. An alternative approach is to integrate the modules much more closely. Invariably such systems are based on an existing drafting system and take the form of optional plug-in modules that provide design or detail oriented functions. The advantage of such modules is that they make use of the interactive graphics capability and the database of the host drafting system and thus appear to the user as a coherent whole.

29. Many vendors have provided reinforcement detailing modules. While these may not alter the suitability of the underlying drafting system for construction applications, a good module will substantially increase productivity when detailing. Features provided include the automatic generation of bar schedules and checking against inbuilt detailing rules. Detailing can be performed with 2D systems but 3D systems clearly have advantages, particularly when reinforcement is congested. Since existing concrete outlines are reused in the detail drawings, a detailing module can help justify the

use of computer drafting.

30. Impressive systems have also been developed for the design and detailing of structural steelwork but these are generally based on the more expensive 3D and GSM drafting systems. Several originated in the process plant sector where high computing costs can be more easily justified. However the most significant factor is that structural steelwork is fabricated in factory conditions and production is being progressively automated. Thus fabricators are interested in computerised design both as an aid to tendering and as a source of accurate data for the production process. With the exception of pre-cast work, equivalent advantages do not exist for reinforced concrete.

Specialist Systems

31. A number of software vendors have produced specialist integrated systems directed at vertical-markets, particularly the design of steel portal frames (Ref. 10). Freed from the need to provide the degree of flexibility required by general systems, such specialist systems have had considerable success.

FUTURE DEVELOPMENTS

32. Existing software packages will continue to develop but there is considerable scope for the introduction of new ideas and techniques. For example at Leeds University we are developing the idea of "Calculation Processing", the manipulation of design calculations in a way analogous to word processing.

33. The most influential of the new techniques are likely to be based on Artificial Intelligence, particularly in the form of Expert Systems. Applied research has expanded rapidly and the results will soon start to be reflected in production tools. Early systems will include diagnostic, specialist advice and training modules but the long term impact on design/detailing will be considerable (Ref. 11). Potential exists not only for improving the engineering aspects of software but also the flexibility and efficiency with which it can be used.

34. Before computers become a normal production tool a more capable and flexible means of integrating individual design tools is needed. It is likely we will see the emergence of general engineering project databases that are independent of specific applications and able to interface to a wide range of design tools. The forthcoming Eureka research project on the integrated design and manufacture structural steelwork (Ref. 12) will probably incorporate this concept.

CONCLUSION

35. Commercial pressures on civil engineers to embrace the computer as a production tool for design and detailing will steadily increase. Good software already exists but does not yet offer the degree of flexibility and efficiency that less committed users will demand. The introduction of Artificial

Intelligence and engineering database techniques will lead to flexibile integrated systems that will give the engineer greater confidence in the software and will address communication problems within the industry.

REFERENCES

1. HOLGATE A. The art in structural design. Clarendon Press, Oxford, 1986.

2. MAHER M.L. and FENVES S.J. HI-RISE: An expert system for the preliminary structural design of high rise buildings. Knowledge engineering in computer-aided design, Ed. Gero J., IFIP, September, 1984.

3. HOGLEY J.R. and KORNCOFF A.R. Artificial intelligence in engineering: A revolutionary change. Applications of artificial intelligence in engineering problems. Ed. Sriram D. and Adey R., Computational Mechanics Publications, 1986, 1155-1160.

4. HOULTON J.M. MOSS integrated modelling, design and drafting for civil engineering. CADCAM '86 Conference, EMAP, London, 1986.

5. TAIG I.C. Expert aids to finite element system applications. Applications of artificial intelligence in engineering problems, Ed. Sriram D. and Adey, R., Computational Mechanics Publications, 1986, 759-770.

6. CAD system sales survey. Construction Industry Computing Association, Cambridge, 1986.

7. WATSON A.S. The impact of computer drafting on the civil engineer. CADCAM '86 Conference, EMAP, London, 1986.

8. BLOOR M.S. et al. Computer aided design interchange of data: Guidelines for the use of IGES. National Economic Development Office, 1984.

9. ALCOCK D.G. and SHEARING B.H. GENESYS - an attempt to rationalise the use of computers in structural engineering. The Structural Engineer, 1970, Vol. 48, April 143-152.

10. YEADON R.E. The interactive design of protal frame buildings. The Structural Engineer, 1979, Vol. 57A, September, 188-292.

11. ROSENHAM M.A. et al. Expert systems applications in computer-aided design. Computer-Aided Design, 1986, Vol. 18, December, 546-551.

12. YEADON R.E. European initiative for producing complete integrated manufacturing systems for structural steelwork Computer systems for detailing reinforced concrete and structural steelwork, ICE, 1987.

Paper 8. The 3D approach to steelwork design and detailing

R.J. ASHWORTH, BSc, MICE, Intergraph (Great Britain) Limited

This paper will principally illustrate the approach to Steelwork Design and Detailing task adopted by the Intergraph Corporation. It will briefly address the history and reasoning behind the approach as opposed to the traditional 2D treatment.

The major portion of the presentation will be to review the modelling and analysis/design software which enables the easy creation of complete 3D model of a structure. Consideration will also be given to the implementation of International Codes of Practice, in particular BS5950 for the design process.

THE INTEGRATED STRUCTURAL SOLUTION.

1. A number of years ago Intergraph developed a system of industrial plant engineering software packages. This enables the plant design team to have a set of tools available for the design and detailing of a whole site, from the initial schematic layout through 3D modelling of equipment, piping and structure, to the creation of the final working drawings and schedules. Originally the structural element that was integrated within this suite of software was targeted towards the type of structure used for the support of vessels or pipework, and other associated structures.

2. The original suite of programs are now in their second generation and the users of the software have also modified their requirements. If the case of the structural area is studied it is found that there has been a shift from the pure plant design type of structure to one of greater variety. Often on a site there will be office accommodation for the permanent administration which may well be designed by an architect. If the

structure is in the oil production area there may be accommodation modules to design in conjunction with the jacket structure. Jackets in themselves were a requirement requested by many users. Architects and engineers are making use of the 3D structural package for visualisations of projects for client presentations and tender documentation. The ability to quickly model, visualise, quantify and cost a new project is very desirable.

3. It is in this area of greater diversity of use of the software that Intergraph has provided a set of tools that are not targeted only at the structural engineer, but at anyone who has a need to create a structure, be they architect or piping engineer. However in making the packages easy to use by any discipline it is not intended to replace the role of the structural engineer. It is unlikely that his position contractually will be affected.

4. An example of how this system can work is given by one of our users. The structural department of the company supplies a very basic set of available steel sections ie. a heavy and a light beam, a heavy and light column. The piping engineer can then work from his preliminary site layout in order to "grid out" the position and size of the building. He can then create a structural layout for his equipment and pipework, safe in the knowledge that any member that he places will be adequate structurally for the job. This then very quickly establishes the project and when the layout is confirmed the structural department is then able to take the structure and add the bracing and loading.

5. In the example above the workflow is speeded up because the piping engineer is not detained waiting for the final structural design. However as the structure is designed, changes will be made to section sizes and bracing will be added to suit lateral forces. Here the other disciplines need to be kept up to date on the progress of the structural engineer. Intergraph provide the capability of Reference Files. This is the ability for an engineer in one discipline to attach, in read only form, the design file of another discipline. In this way he is able to review the changes as they are made and act accordingly. It is important that the file attachment is for reference only. It is inappropriate for the piping engineer to have access to the structural database or calculations programs.

6. Due to the developments in computer communications the use of reference files is not restricted to one site. A company in Reading performs its piping and structural design with reference to site works and drainage files in Glasgow. For one particular project a trans-Atlantic link was made.

7. When the time comes for the creation of contract drawings, reference files can also be used. The site layout file can be referenced to the drainage and earthworks models for the creation of the foundation drawing. Only those files relevant to the task need to be displayed to convey the information to the site engineer concerned. The same files may be used in a different combination for the road layout drawing. There are extensive techniques available for the use of reference files, including the ability to attach a file to itself as a reference file. This gives the ability to automatically update a drawing when a modification is made. For example, if the top of steel level is changed in the plan view the elevation is instantly changed. This is a technique that is only available in 3D.

THE STRUCTURAL ENVIRONMENT

8. This paper is only concerned with the 3D products that Intergraph provide. However there are a number of 2D products which are concerned mainly with steel and concrete framed structures. It is important to note that these packages, although they provide 2D graphics files, have a full 3D alphanumeric database behind them. For example when a column is placed on a floor level in the 2D file the operator has to tell the system where the column starts and ends. He does this by defining a parametric column the basic attributes of which are; section size, material grade, base elevation and splice elevation. This then enables the operator to obtain a material report for the project. There is also another use for the elevation or "Z" information.

9. All Intergraph structural products have a read/write interface to the "Common Structural Database"(CSD). This is a neutral, published file format which acts as the medium between products. A particular use for this would be to convert a 2D model into a 3D model. A simple example of this would be for an offshore jacket structure. The

topside modules are basically orthogonal in layout, and the floor levels can be quickly laid out. The jacket structure will be modelled in the 3D package as the geometry of the legs and braces are more complex. Also there is the capability for the detailing of the tubular intersections, and the "flat pattern developments" of the tubular members. At some stage the overall centre of gravity of the structure needs to be established. The 2D model of the module can be converted into a 3D model via the "CSD" and placed on top of the jacket. The centre of gravity of the whole structure can now be calculated.

10. Similarly if a 3D model of, say, an office structure is produced the "CSD" can be used to read the model into a 2D drawing file floor by floor for the issue of the drawings. The "CSD" can also be used as an interface to other software. If an analysis deck is produced from Third Party software, it can be re-arranged into the "CSD" format and then read into any of the structural products. This again eliminates multiple entry of data.

3D STRUCTURAL MODELLING

11. A primary objective of the writing of the 3D package was its ease of use. It was realised that there may be resistance from new and existing users to a 3D approach. Also there was the feedback from users of the original software. Based on this the package now provides the tools for the experienced 3D users to make much faster progress, but gives the new 3D user tools that will be familiar to him from a 2D world. Typically these include the ability to work on a floor by floor basis and then work on the elevations. Extensive use can be made of the grid layout system which may have been generated in, say, an architectural design file.

12. Another important tool that is made available is that of member associativity. This gives the modeller the ability to move an individual member, or group of members, and all associated members also move. Typically a whole floor can be moved vertically to suit a new service void requirement, or the end bay of a structure can be moved laterally to accomodate a new width of side door. This is an interactive function and the database behind the graphics keeps track of all the member length calculations. A more advanced use of

this technique was the modelling of a satellite communications dish. Here the geometry of the dish was set out in plan as a typical of a dish series. Then the dish can be associatively moved onto the correct parabolic curve in plan. If subsequently it is required to change the radius of the dish this also can be done associatively.

13. A further important tool for the effective modelling of offshore structures is the inclusion of "rule based" joints. The rules are based on those described by the American Petroleum Institute in order to offset member centre lines at a joint to enable weld clearances to be obtained. The package is able to calculate the positions of the external faces of the members and then can offset them to give the desired weld clearances. This command is able to work in 3D so that out of plane members can be catered for.

14. When the member cutback calculations are performed the system is able to give a material report based on the cut length of a member or the material length. This enables different reports to be generated for materials ordering and for fabrication. Also centre of gravity and centre of buoyancy reports can be made.

15. A further feature of the software is the ability to create a "CSD" on part of a model. This enables the creation of a library of structural "modules" to be made. These can then be read into a new model as required. For example the cross-arms of a transmission tower can be kept in a library. When a new tower type is required it can be built up from a kit of parametric parts.

16. At any time during the creation of the model there is the facility to perform "clash detection" between structural elements themselves, or with other elements such as pipes, ladders, access ways. This is done as a batch process on the whole project or more realistically on a part of the building that is known to be congested. The clash report can be in the form of a report or as a design file with the clash markers displayed. This then can be attached to the original design file as a reference file.

STRUCTURAL ANALYSIS AND DESIGN

17. When the "physical" model is complete it can then be passed to the analysis and design

software. Again in order to eliminate multiple entry of data all of the geometry and section properties of the model can be passed through the "CSD" to the Finite Element Modeller. Essentially all that needs to be done to the model is to load and constrain it. The "CSD" reader in the analysis software will insert nodes at each intersection of members. It will split continuous columns into separate lifts in order to satisfy FE requirements, but it will also note in the database that when the column is to be designed it is to be treated as a single physical element.

18. All loads, nodal, UDL, edge or moment can be displayed graphically in the design file that the database creates. In this way a set of illustrated design calculations can be easily produced for the verification procedure. In the same way, fully fixed, pinned, gap, hook, spring and damper constraints can be displayed. This makes for quick visual checking of the structure prior to the analysis.

19. The analysis engine provided allows for static or dynamic solution of the structure. Structural non-linearity can be simulated using gap, hook, spring and damper elements. All of the analysis results are posted back to the original FE database for display of the results. Again graphics play a large part in that shear force, bending , moment and deflection diagrams can be produced, as well as stress contours and animations of deflection cases or mode shapes. The nodal deflections can be posted back to the original model in order to re-analyse making a P-delta approximation.

20. The design processor is also able to use the same database created for the analysis. All that has to be done is set pointers to the required design code tables or equations. Additionally there are facilities to group members together into "mark groups". A mark group is a set of elements that are to have a common design such as a group of floor beams that have similar lengths and loading. The mark group design process will select the same sections for all members in a mark group ensuring adequacy of the section for all members.

21. It is the policy of Intergraph to provide design tools that are of use throughout the world and it contains options for design to: BS5950, DIN, AISC78 for steework. The same product will

also design to the same national codes for reinforced concrete. This has been achieved by extensive work on making the database very flexible in its structure. This allows fast introduction of new codes of practice. Also the same structural model can be designed under each code. Again the verification against alternative codes is of interest.

22. During the set-up for the design process there are a large number of checks that can catered for and various look-up tables that can be established. Typically these are alternate section tables, deflection limit tables, code tolerances, composite section design criteria, alternate shear bolt tables. Also section design criteria can be set. For example, beams can be limited to a maximum or minimum depth or weight. An interesting illustration of the type is the treatment of the bracing parameters for design elements.

23. The bracing parameters can be specified for each element or groups of elements in a model. These parameters include the unbraced lengths for bending about the major and minor axes. The bracing for strong axis bending has been further separated into bracing for the top and bottom flanges of a section. This allows for the accurate description of the bracing for both the positive and negative moment regions for continuous beams and rigidly connected frames. The warping constraint at member ends can also be entered for models with open sections subject to torsional loads. Also included is a flag to indicate whether the members are braced against sidesway about the major and minor axes.

24. When a design has been processed the results of the design and the modified sections sizes can be posted back to the analysis engine for re-analysis. The analysis/design cycle can then be repeated until the design reaches convergence. Full reporting facilities are available on the calculations in the format of the code required.

25. When the design is complete the results and new sections sizes can be posted back to the physical model via the "CSD". This then completes the cycle. Drawings and schedules can be extracted as required.

CONCLUSION

26. With its nucleus products of powerful graphics and database, Intergraph are able to present the engineer with very powerful tools for the analysis, design and detailing of highly complex steelwork structures. In addition the capability to make complete material lists and calculations has proved to be most productive. The principal strengths are the ability to integrate many disciplines and to shorten project communication routes. The ability to detect clashes before a project reaches site is also proving to be very desirable to client and engineer alike.

Paper 9. European initiative for producing complete integrated manufacturing systems for structural steelwork

R.E. YEADON, BTech, PhD, MIStructE, FBIM, Computer Services Consultants (UK) Ltd

SYNOPSIS. The Paper will describe a European initiative for producing complete integrated manufacturing systems for structural steelwork. Many of the elemental components for integrated manufacturing systems have been addressed such as design, analysis, estimating and structural detailing. It is apparent that the most successful areas for computerisation have been in the structural analysis and design areas. Over the last 15 years a number of attempts have been made to produce automatic detailing systems for structural steelwork. The Paper will discuss the philosophical and technical strategies adopted and review the future development requirements in the context of the new Project.

INTRODUCTION

1. Over the last 15 years many attempts have been made to produce structural steelwork design, analysis, drafting, detailing and estimating systems. The early attempts were based around large, main frame, expensive, computer systems using batch mode oriented techniques.

2. Possibly the earliest significant package was produced in Canada by Task Master Systems for the comprehensive analysis and design of steel frame buildings. The system required the description of the building in a co-ordinate and connective form - load information, support conditions etc. The system would then produce designs and design output for the complete building including connection forces etc. For various reasons this approach has not been totally successful. The input techniques adopted were batch mode oriented. In consequence - input could be tedious, errors were difficult to trace and correct, and the systems were generally unwieldy and relatively expensive.

3. The advent of the CPM 8 Bit Micro Computers heralded the possibility of mass computing. However the early implementation of these systems did not realise the true potential - lack of standardisation of Operating Systems and Languages and the relatively low power of the systems inhibited the development of software for comprehensive

engineering applications. Numerous design modules for elements were produced and proved very effective on this type of machine. However integrated analysis design and detail systems were not practical or feasible.

4. More recently the entry of the IBM PC and its compatibles together with Unix 32 workstations into the marketplace have realised a number of requirements for the effective implementation of comprehensive engineering systems. These are

(a) relatively low cost
(b) significant power
(c) standardisation of operating systems and languages

5. For a number of years The British Constructional Steelwork Association has been active in encouraging the development of computer aids for steelwork design and detailing. During the late 1970s and early 1980s a high level of activity has been evident in the study of the market and the steering of developments in the structural steelwork sector. In 1980 - Ove Arup was jointly commissioned by the BCSA, BSC and PSA to carry out a review of the structural steel industry and the requirements for computer aids. At that stage it was concluded that computer aids were of a lower priority in terms of development in a number of other areas.

6. Increasing the competitiveness of the fabrication sector with the introduction of more efficient working practices and shop floor tools/techniques was perceived as an essential part of expanding the structural steel market. More effective marketing of steel products and cost-effective solutions for multi storey buildings were perceived as essential pre-requisites of a successful structural steel sector. The manufacturing industry has gone through a period of decline and in certain sectors it is apparent that a leaner, fitter industry is now appearing. Overseas competition has however been a significant factor in this decline and it has now become apparent that the structural steel sector is being threatened by such competition. Structural steel Projects are pre-fabricated off-site and then transported to site for erection. Transport costs are relatively low and therefore such pre-fabrication could be undertaken as far away as Japan and transported to Europe for erection. In early-1986 Nippon Steel carried out a technical appraisal of computer aided design systems available in Europe. It was declared at this time that the Japanese Industry was in the process of investigating computer integrated manufacturing to improve the efficiency and competitiveness of the Japanese fabricated steel industry. As a result of a successful Japanese CIM initiative - significant imports of fabricated constructional steelwork into Europe would be evident. Significant surplus capacity in the Japanese

fabrication industry is already evident and only a relatively small proportion of this would need to be diverted into Europe to have a devastating effect upon the indigenous industries.

7. Against this background a European initiative was conceived in mid-1986 to produce a computer integrated manufacturing system for constructional steelwork on a pan European scale.

PROJECT DESCRIPTION

8. The objective of the Project is to undertake a research study into a modular computer integrated manufacturing system built around a product modeller for building structures that consists of a number of sub-systems for computer aided design as well as computer aided manufacturing for the steel fabrication industry. The system will consist of sub-systems for:

Design

9. Interactive design system for steel structures both for multi-storey buildings, low rise buildings, bridges etc. The loads can be given interactively or based on national codes of practice or the new European Commission's "Eurocodes". The results are presented alpha-numerically for presentation to approval bodies and in scheme drawings. In addition the generated information is retained in the product model.

Analysis

10. The information from the design phase is retrieved from the product model and formatted in such a way that it can be used immediately in the existing framework analysis programs.

Cost Estimation

11. The primitives, the product model is built-up with, also contain information about the fabrication processes which are needed to realise the required geometry. After the detailing is finished, all project information that is needed for cost estimation is also available for the material list and for calculating the necessary man hours.

12. Project independent files with times for fabrication processes make a detailed cost estimation possible.

Detailing

13. Detailing comprises design and calculation of the connections and modifying the design of the total structure for possible architectural provisions.

Work Planning and Fabrication

14. For fabrication, working drawings, bills of materials, order lists, programming of NC machines and welding robots - material resource planning is needed.

Country	Collaborator	FINANCIAL CONTRIBUTION Phase I (£000)	QUALITY ASSURANCE [QA/QC Method Statement Standard(s) Rationalisation]	PRODUCTION [Production Cells, Robots, Cell Layout, Simulation, Erection]	PRODUCTION PLANNING [Estimating, Management Information Systems]	PRODUCT DESIGN [Product Modeller, Design, Analysis, Estimating, User Interfaces, Detail Design, Connection Design, Prefabrication Innovation]
Great Britain	BCSA		x			x
	CSC		x			x
	TWI			x	x	
	AMTRI				x	
	ROBERT WATSON				x	x
	NUSTEEL				x	x
	BILLINGTONS			x	x	x
	WARD GROUP					x
	LEEDS UNI					x
	SHEFFIELD UNI		x			x
	OVE ARUP		x	x	x	x
Netherlands	IBBC-TNO		x		x	x
	MI-TNO			x		
	H EVERS			x		x
	MOESKOPS BV					x
	HOLLANDIA KLOOS					x
	DELFT-UNI					x
	BRANCHEGROEP STAALBOUW		•			
	STAALBOUWKUNDIG GENOOTSCHAP					
EC	EUROPEAN COMMISSION (DG III)	*				x
France	CTICM			x		x
	CGA-HBS (ALCATEL)			x	x	x
	CRIIF-(LIMSI)					x
	ETS NOUVEAU JOUFFRIAU		x			x
	RENAULT-AUTOMATION			x		x
	INSTITUT DE SOUDURE			x		
Italy	ITALSIEL	*	x	x	x	x
	SIDERCAD	*				x
FRG	BOCAD	*	x			x
	BOCUM	*	x			x
Finland	JUVA OY			x		x
	PPTH OY					
	CADEX OY		x			x
Denmark	RAMBOLL & HANNEMANN		x	x	x	x
	SVC		x	x		
	INTERNATIONAL STEEL CONSULTING					x

* Still under discussion

Fig. 1. EUREKA Project 130 - Computer integrated manufacturing for constructional steelwork, including expert systems matrix of collaborator involvement

15. A detailed study will be undertaken of the fabrication methods currently employed by the industry together with an examination of the potential for the application of robotics (together with their possible influence upon structural forms). "Expert Systems" will be evaluated with regard to their possible application to optimising new and existing factory layouts and configurations.

PARTICIPANTS

16. The Project will be led and co-ordinated by The British Constructional Steelwork Association (BCSA). Thirty-six Organisations are involved as Participants from seven European countries. All of the participants, together with Governments - will contribute through the provision of funds, research effort, equipment, practical experience and expertise. A Project Organisation Chart is given in Fig. 1.

17. European participation will be directed via the European Convention for Constructional Steelwork - of which the BCSA is the United Kingdom Representative. The other European Representatives consist of the BCSA's sister Organisations throughout Europe. The ECCS has a full-time Secretariat based in Brussels.

18. The two major UK Collaborators will be The Welding Institute and Computer Services Consultants Ltd. Two major steel fabrication Companies will act as "test sites" for the Project.

RELATIONSHIPS TO OTHER EUROPEAN DEVELOPMENTS

19. In addition to being based upon national standards, the Project will be based upon the European Commission's emerging EUROCODE 3 for the design and fabrication of steel structures. The Commission intends that EUROCODE 3 will establish a set of common rules as an alternative to the differing rules in force in the various Member States. This falls within the Commission's programme for aligning the regulations, laws and administrative provisions of the Member States concerning the safety, serviceability and durability of the different types of construction and materials.

20. The objectives of the EUROCODES are to:

(a) promote functioning of the Common Market by removing obstacles arising from differing rules
(b) provide common technical rules for an efficient application of the Council Directive 71/305 on the co-ordination of procedures for the award of public contracts, which can be applied as an alternative to the national rules.
(c) reinforce the competitive position of the European Construction Industry and associated industries and

activities in Europe in relation to countries outside the Community

(d) establish a harmonised basis for the intended common rules for building products.

21. The EUROCODES will provide an optional set of design rules which can be applied within the Community as an alternative to the corresponding national rules covering the same technical matters.

22. The safety level of structures will be defined by each Member State in accordance with its national requirements. In order to permit the application of the common rules, each Member State will determine appropriate numerical values for the partial safety factors and the other safety elements. The application of the EUROCODES and the continuation of the harmonisation effort will permit the provision of the gradual establishment of common values.

23. The control of design and execution and any approval procedure of structures will remain subject to national regulations. The same applies to technical supplements with regard to aspects which are not yet comprehensively covered by the EUROCODES or which cannot be covered in terms of generally applicable rules.

24. The unification of the European markets for constructional steelwork will be significantly assisted by the advent of Eurocodes and the harmonisation requirements embodied in the EEC objectives. It is envisaged that the Project being undertaken will assist in cementing the currently disparate market and will provide a focus for a common European Market development. The initial considerations for system evaluation will be based upon developments to date undertaken by a number of the Collaborators - Product Modellers from Bocad of Germany and Renault Automation in France; Design and Analysis Systems from CSC in the UK, TNO in Holland, Sidercad in Italy and Cadex of Finland. The offerings will be evaluated with regard to the Project objectives. In addition a number of other Systems will be considered as the appropriate vehicle for the CIM initiative and the first phase of the Project will be defining these vehicles and system architecture.

25. In conclusion the limitations which have constrained systems development in the past such as availability of hardware/software systems, resources and market potential - now appear to be overcome and it is considered that this pan European initiative to develop computer integrated manufacturing systems for the structural steel industry will be a milestone in the harmonisation of the European marketplace and present the relevant response to the commercial and economic pressures currently being imposed.

Paper 10. Structural steel detailing and scheduling: PC and workstation solutions

G.P.P. RIDE, Radan Computational Ltd

SYNOPSIS. Radan Computational Ltd. have recently introduced a Structural Steelwork Detailing and Scheduling applications program which is able to run with two CAD drafting packages. Alongside Autocad it satisfies the popular PC market and with Radan's own Radraft it is available to the more powerful multi-tasking 32 bit UNIX workstations such as SUN and Whitechapel. The problems faced by the Steelwork Fabrication industry and the contribution made by SSDS towards solving some of these are described. In addition the appropriateness of the PC or workstation solution to particular situations is considered.

THE PROBLEM

1. There is a heavy design, detailing and scheduling load on Fabrication Companies in the UK because of the structure of the Industry. In the reinforced concrete structure the Structural Engineer is generally responsible for both RC detailing and Bar Bending schedules but this is not the case for steelwork where the fabricator is responsible for converting design parameters into a finished structure.

2. This process (see fig.1) is both complex and time-consuming, the quantity of scheduling which has to be done is particularly onerous and prone to errors. To make matters worse for those prepared to solve these problems, through the medium of CAD, fabricators have, naturally, adopted their own varied methods of coping with the difficulties over the years. This additional layer of "traditional Company method" makes fig.1 only an example and not a definitive process. Different jobs may also force a Company to by-pass or alter elements in their own traditional approach.

3. Since virtually all contracts are won on price the Steel Fabricator has to be constantly vigilant with respect to his costs. Discounting the obviously foolhardy approach of cutting design corners beyond the limits of morality ther remain three main areas of savings which can be considered. These are, shop fabrication, site erection and drawing office.

4. The recession in the Construction Industry over recent years has undoubtedly left the industry with only the most efficient survivors. It is probably true that in most cases the fabrication shop is unlikely to be able to yield up very much

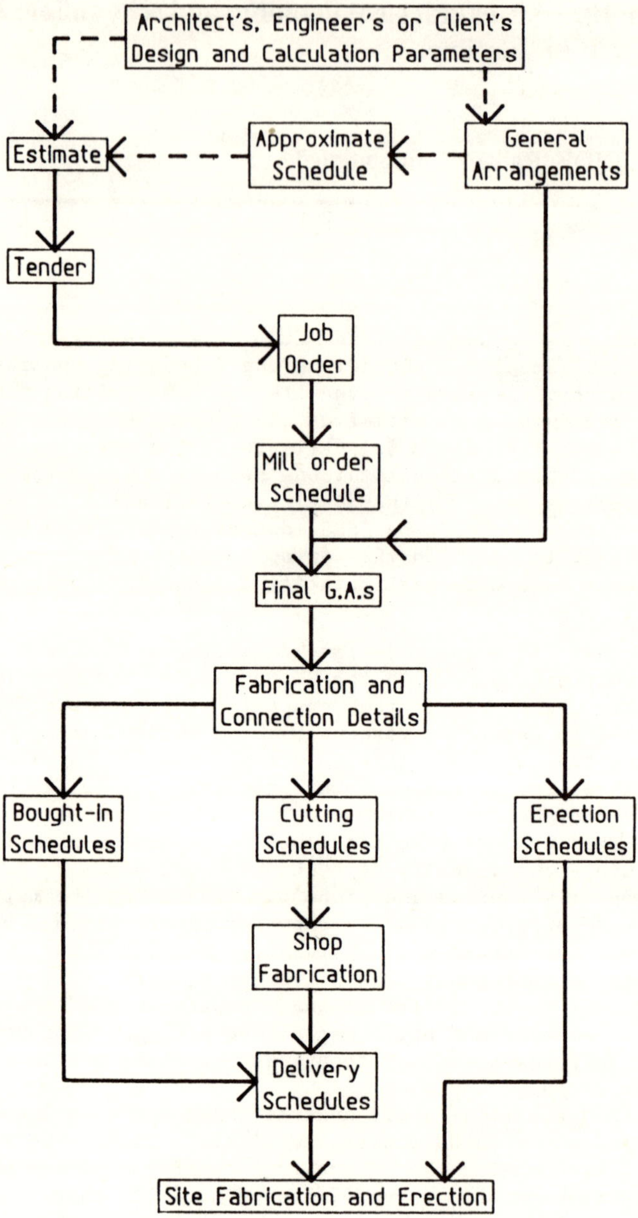

Fig. 1. Structural steel building process

in the way of further efficiencies and the vagaries of the British climate make it extremely difficult to apply a consistently efficient approach to the typical UK building site.

5. However, it is my contention that savings within the drawing office of significant magnitude are possible by the use of a specific steelwork drafting package. The Fabricator's drawing office is undoubtedly the "Engine Room" of the Company since, not only do General Arrangements, Fabrication and Connection Details emanate from it but Buying and Cutting Schedules are also produced. Taking on some of the problems faced by this sector of the steel industry and solving them has become possible by the introduction of considerably more cost-effective computing hardware over recent years. This has enabled software application houses like Radan Computational to apply themselves to this specific task in a meaningful way.

THE BENEFITS

6. An integrated steelwork detailing and scheduling package based upon a fast drafting package with a 200 to 400% increase in output can pay for itself within 9 to 18 months now that the cost of computing power has dropped.

7. In addition to this greater throughput, earlier and more accurate schedules of materials are possible thus assisting in the resolution of the current trend towards fast track schemes. The final benefit of a specific CAD package comes from the accuracy of the drawing and the associative dimensions created. This benefit is taken up unconciously by the fabrication shop and relayed to the site erection team whose realisation that improvements have taken place is exemplified by the ease of erection and the decreased use of the burning torch.

THE SOLUTION

8. The advent of PC and 32 bit UNIX workstation hardware at lower costs referred to in paragraph 5. means that an affordable and cost-effective applications package is an achievable goal. In order to tackle the requirements of a solution effectively a proper understanding of both, computer aided drafting and its interaction with databases and the Fabricator's needs is vital.

9. The remainder of this paper describes in some detail the application program SSDS which runs alongside Autocad for PC based systems or integrally with Radraft on workstations. Differences between the two versions are born out of the different hardware potentials and the functionalities of the host CAD packages. These are dealt with at the end of the paper in order to allow a description of the generalities without constant comment on the variations.

10. The basic ingredients of a CAD solution to steelwork are:

a) An efficient and fast 2D drafting system with a high degree of functionality.

b) A database of the available Steel Sections to BS 4:Part 1.

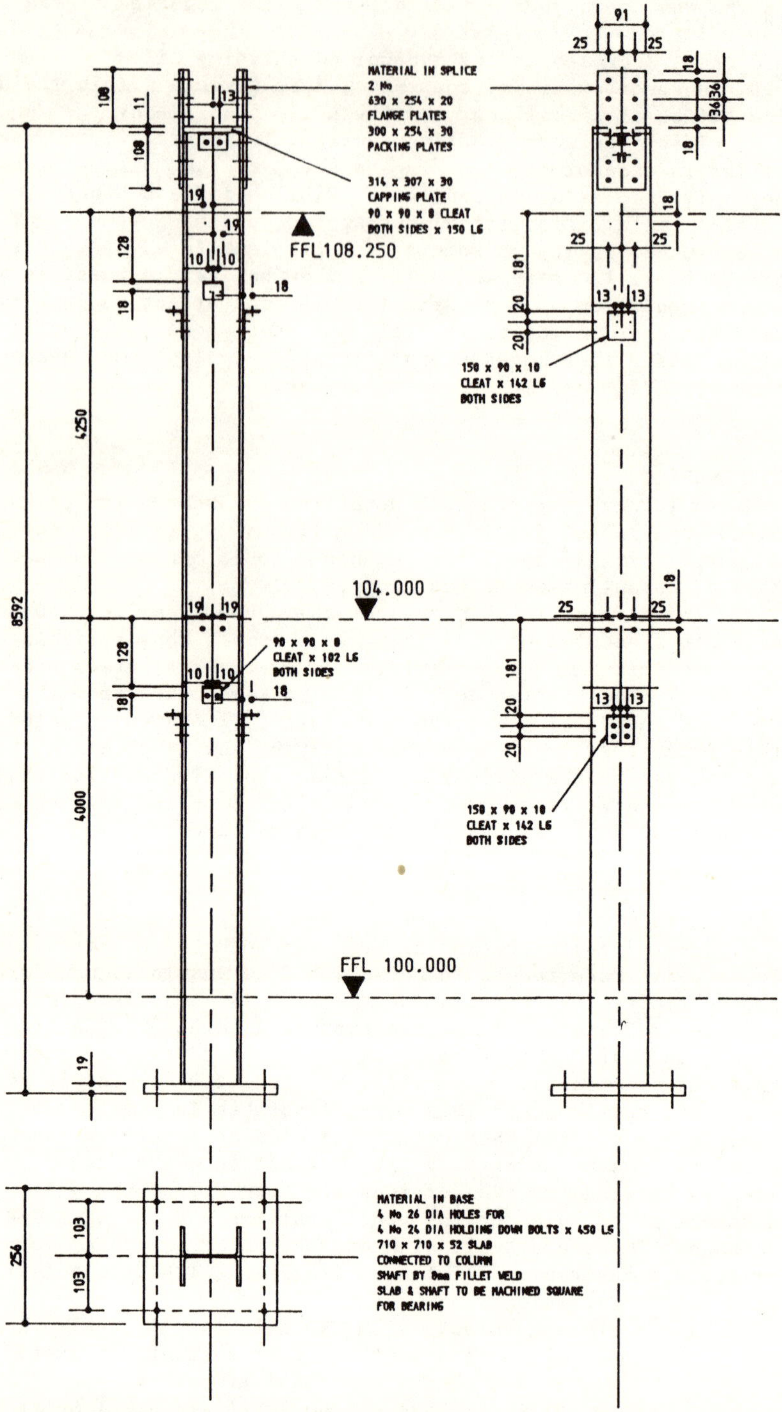

Fig. 2. Typical Column Detail

c) An integrated means of scheduling sections which have been incorporated into the drawing
d) An easy to use interface with the steelwork detailing operating the CAD system.

11. Both Autocad and Radraft satisfy the requirements of a) within the confines of the hardware upon which they operate. Radan's SSDS software provides the solutions to the remaining ingredients although there are major differences between the presentation of the user interface due to variations in overall concept between the host systems.

DATABASE

12. Typical steelwork G.A's and details contain graphical representations of various steels, connections, bolts and welds (see fig.2). Standard steel sections are available from steel mills in accordance with BS 4: Part 1 for Universal Beams, Columns, Channels, Angles, Hollow Sections etc. These are specified as a nominal size plus a mass per metre. The actual overall size and web, flange and wall thicknesses differ, of course, depending on the mass. The difference between nominal and actual sizes is critical to the lengths of connecting members, type of connections used and the co-ordination of the building components which will hang on the steel skeleton.

13. SSDS contains a database of these standard sections and the specification of a particular section within SSDS enables a display of any of five views of the steelwork, top, bottom, sides and cross-section, to be incorporated into the CAD drawing. In addition to the fact that this represents some 2500 to 3000 separate possibilities is the need to portray these sections on drawings with any desired length from the actual total length to any convenient portion for the detail in question. The ends of sections may also need to be shown closed, notched, open or as break lines. Taking all these factors into consideration it is clear that a parametric approach, rather than a symbolic one is vital. This is, in fact, how SSDS solves the problem.

14. As the user specifies the steelwork required within SSDS the actual length needed for the section is scheduled into the Bill of Materials singly or in multiples dependent upon the user. The production of both the graphical display and the scheduling of the steel in one action is clearly economic from the detailer's point of view.

USER INTERFACE

15. If, for example, the user has to type in 305x127x37 UB in order to invoke the CAD system then structural steel detailers will be involved in unneccessary typing. Therefore the SSDS user interface has been designed to enable the detailer to select through either a form or a series of questions steel types, sizes and masses merely by number (see fig.3 for example). This kind of presentation is friendly, easy-to-use and enhances productivity.

INTEGRATED DETAILED DRAFTING

16. Having established the primary components of an available Steel Sections Database and a good user interface to interact through, the next stage is to use this to greater benefit by utilising the central database as a kind of model of the steel which is being input. The SSDS integrated database allows the detailer to instantly create the graphics for any of the available views or the annotation of particular

```
Current Field :     Bottom cope dim                          SELECT ON

                   DETAILS OF NEW MEMBER : SIZE

Reference Name  [beam            ]  End #1 (C/P)       [C]
Assembly_mark   [b-]                   Top cope dim    [2-1/2 x 3/4   ]
Steel Type      [W ]                   Bottom cope dim [              ]
Designation     [1 ]                End #2 (C/P)       [P]
No. off         [1 ]                   Top cope dim    [            > ]
Grid Length     [1'        ]           Bottom cope dim [              ]

  ( 1) 2-1/2 x 3/4                  ( 7) 4-1/2 x 1-1/4
  ( 2) 3-1/2 x 1                    ( 8) 4-3/4 x 1-1/2
  ( 3) 3-1/2 x 1-1/4                ( 9) 5-1/2 x 1-1/4
  ( 4) 4     x 1                    (10) 5-1/2 x 1-3/4
  ( 5) 4-1/4 x 1-1/2                (11) 5-3/4 x 1-3/4
  ( 6) 4-1/4 x 1-1/2                (12) 6     x 1-1/2
```

Fig.3. "Add Steel" Menu - Autocad version, AISC Database

steelwork merely by identifying the steel already represented on the drawing. This action does not affect the content of the Bill of Materials and therefore is strictly for displaying other views of steelwork and not for adding it to the schedule.

17. Connection details are either bolted or welded by using angles and plates in various combinations of type, size and orientation. SSDS enables the introduction of such connections by commands such as:

Column Splice Plate
Stiffened Plate Connection
etc.

18. In conjunction with the above, typical bolting grids (3x2, 2x2 etc.) may be input together with recommended plate or angle sizes. The selection process is simple and produces drafted connection details with the minimum of input (see fig.4).

19. The internal database recognises each physical structural entity in addition to its graphic representation. As with main steel sections, connection details defined in one view are automatically shown when corresponding views are requested. This associativity even extends to bolt holes through cleats and steel webs.

20. Some "intelligent" dimensioning routines have been provided in accordance with industry standards. Dimensioning is a complex and demanding process and since it informs the fabrication shop of the correlaton of components and holes it is of vital importance. Precision is required so that there is no mis-match between shop produced items when they are hoisted into place during the erection process.

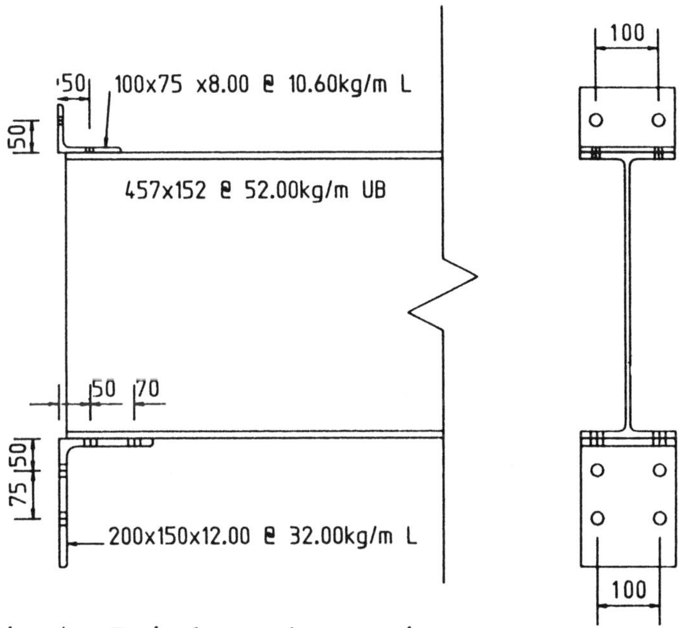

Fig. 4. Typical seated connection

21. Take an example of a framed connection. This complex process of dimensioning consists of:-

a) Bolt hole spacing
b) Bolt hole centre from back of angle
c) Bolt hole to bolt hole of adjoining angle
d) Back to back angle spacing
e) Bolt hole from edges of steel.

22. Not only can SSDS cope with the precision required but this exercise is completed by SSDS with a single command (see fig.5).

23. Welding notes and graphics can be generated from a library of industry welding practices. A form-filling approach defines the type of welding use in the case of the Autocad version. A symbolic approach is taken in the Radraft version whereby the CAD system is used to indicate the location of a weld and then the appropriate weld type is called back from a standard library.

24. A Bill of Materials can be printed out at any time and this quantifies the numbers, lengths and weights of the various steel sections and bolts on the drawing. Manual production of

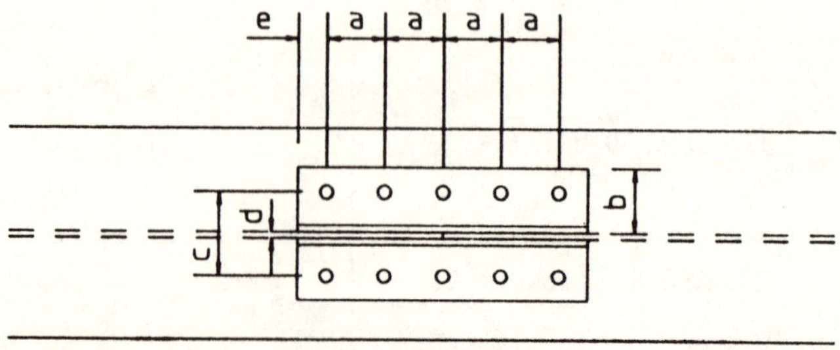

Fig. 5. Hole dimensioning

Site ref: Hyperstore GA/1.b Bill of Materials ref: 8745/1

BILL OF MATERIALS

QTY	SERIAL SIZE	LENGTH (mm)	WEIGHT (kg)	COST (£)	REM
12	457x152x52 kg/m UB	3090	1928.16		
12	457x191x67 kg/m UB	8990	7227.96		
6	254x146x31 kg/m UB	2770	515.22		
8	254x254x89 kg/m UC	2770	1972.24		
10	229x89 x32.76 kg/m [	6460	2116.30		
24	125x75 x12.0 L	7455	3184.78		
36	100x75 x10.0 L	150	70.20		
6	100x75 x12.0 L	180	16.63		
16	150x75 x10.0 L	225	61.20		
4	150x100x10.0 RHS	5500	785.40		
8	200x100x6.3 RHS	2350	532.04		
4	150x150x6.3 SHS	2350	266.02		

Fig. 6. Bill of materials output from SSDS

this schedule is one of the most tedious and error-prone processes which the Steel Fabricator has to perform. The time saved on this task alone when coupled with the reliability of the schedule in relation to the drawing can more than justify an investment in SSDS (see fig.6).

25. Because SSDS has an integrated data structure editing of a steelwork component may be individual or associative. Details of sections may be amended or deleted and in the case of the latter all views and annotation disappear from the drawing and schedules. Deletion of cleats or plates will also delete the bolts associated with them.

LIBRARIES OF USER CREATED DETAILS

26. Over a period of time the user would be able to create libraries containing standard situations to the Company which would further increase the productivity of both schedules and drawings. Amendments to components on these standard details would be reflected in an updated schedule just as it would be in the editing procedure previously described.

AUTOCAD/PC AND RADRAFT/WORKSTATION DIFFERENCES

27. The differences between the two available systems are dependent on two principal factors, hardware and software concepts. Software is constrained or otherwise by the operating system of the generation of computers upon which it is designed to be installed. Similarly the speed, power, capacity and efficiency of the application program is dependent upon the computer's specification as well as the programmer's skill. Consideration of hardware and operating system difference will be given first.

HARDWARE

28. In general terms the operating system of a computer is in-built software which allows the programmer to address the computer and which provides utilities enabling such facilities as filing, graphics, window management, peripheral interfacing etc. to be dealt with efficiently by the programmer. PC computers use a system called PC-DOS and the particular workstation used by Radraft is UNIX based. In addition to having more general utilities UNIX is a multi-tasking and networking system whereas PC-DOS is not.

29. Multi-tasking means that the processor and operating system of the workstation have the power to do more than one process at a time. In the context of steelwork detailing this means that the operator can be using the CAD system to do a drawing and at the same time, be plotting out a drawing, printing a Materials Schedule and running a design program. The drafting package in this case will be in foreground and the other processes in background. With the use of the window manager system however, the operator can switch from one process or another to be in foreground. This ability of UNIX workstations obviously increases efficiency and flexibility.

30. Networking of workstations together and to a larger hard disc than is generally available to PC's is the other major difference. Hard discs are typically upwards of 100 Mbytes and

workstation systems usually take a modular approach allowing upgrading of disc space to grow with the user. Unless the software takes advantage of networking abilities nothing is gained, Radraft has networking and this enables full sharing of databases, peripherals etc. without any loss of computing power which resides within each individual networked station.

31. In terms of hardware power PC's are generally slower, have less computing power and smaller hard disc capacity. Workstations incorporate high resolution graphics screens whereas this is an extra to a PC system. Workstations are now competitively priced with PC's particularly when more than one system is required.

SOFTWARE

32. Both Radraft and Autocad have been designed to run on their respective hardware and to thereby fully utilise the hardware potential. The principal differences in the SSDS versions are, cosmetic presentation of the user interface and the abilities of the system to accommodate tailoring to particular needs.

33. With regard to the user interface Autocad SSDS presents a form menu to be filled, whereas Radraft SSDS presents each question individually on a prompt line. This cosmetic variation is merely subject to personal taste.

34. The second difference is more profound. Radraft contains its own macro language, programmable by the user or Radan. This allows fully interactive tailoring of the applications by providing the following facilities:-

a) The chaining of drafting commands
b) The abstraction of geometrical drawing data
c) The writing and reading of files
d) Extensive calculative powers

This enables the Fabricators' method of working to be accounted for and suitable systems to be easily tailored to special needs. It also allows lengths of sections to be automatically calculated from already existing drawing data. File reading and writing allows design program output to be input into Radraft SSDS for automatic section specification as well as the production of additional schedules for say, purlins and cladding. Although Autocad has its version of a macro language it is not possible to use this as extensively with SSDS.

CONCLUSION

35. A viable, cost effective Structural Steelwork Detailing and Scheduling system, SSDS, is available now. This will provide significant savings in the drawing office together with reduction of errors and increases in accuracy.

36. For the small company, the self-employed contract detailer and educational establishments the Autocad SSDS version is ideal. The larger organisation, those with extensive complex drawings or where specific tailoring would be advantageous would benefit considerably from the Radraft SSDS version.

Discussion

MR W. M. YOUNG, Stirling Maynard & Partners, Peterborough
With reference to Paper 4, my company has not yet used its CAD system for this type of detailing, but in my four years' experience as a consultant advising manufacturers on CADCAM under the Department of Trade and Industry Scheme, every application which has shown potential for a large productivity increase has been based on parametric drawings. Could the Author please comment?

MR J. S. L. COOPER, Mid Glamorgan County Council
We are investing heavily in CAD and hope to achieve savings not only in improved staff productivity, but also in reduced final build costs and claims, and in the possibility of loss of grants from government. We also hope to achieve savings over the current costs of our computing systems. Finally, organisations which do not invest in CAD will find it increasingly difficult to attract good quality staff.

Several Papers refer to the computing power needed by CAD systems. At the same time, the problems of managing these computers and their data have also been raised. The modern 32 bit workstation (Apollo, SUN, VAXstation and so on) offers the necessary computing power, together with the ability to share resources and to prevent the misuse of the system. Several large software engineering houses in this country are now confining their developments to work-stations.

Do minicomputers have the power needed by modern CAD systems? Perhaps in ten years' time, minicomputers will be a thing of the past. Could the Authors please comment?

MR T. A. NICHOLSON, Sir Alexander Gibb & Partners
There has been little discussion about the investment costs of training users in CAD. Could the Authors give some idea of how long it takes for detailers to become proficient in the use of CAD reinforced concrete detailing?

CAD SYSTEMS

MR KATTENHORN (Paper 4)
While it is true that using a CAD system as a straightforward draughting tool will increase productivity, it is only when repetitive draughting processes are automated that further appreciable gains in productivity can be achieved.

Parametric symbols have been used extensively within the development of the software for Taylor Woodrow CAD applications. The advantage of the techniques is that development does not require a particularly high level of computing expertise. The generation of libraries of parametric symbols is an extension of the drawing process and could be undertaken by competent draughtsmen.

Until recently, CAD systems were based either around central mainframe/minicomputers with W/S connected to them or on microcomputers networked together. Both of these solutions tended to suffer from the problem of providing adequate computing power to 'keep up' with the draughtsmen at reasonable cost. Improvements in technology have seen the cost of computer power steadily falling, and it is now possible to provide a workstation with the power of the mini at the cost of a micro. Within our organisation, it is envisaged that future expansion of CAD will be 'stand alone' workstations linked to a central file server for drawing storage and security.

Within our organisation, CAD draughtsmen are given an initial two week training course, at the end of which they are normally able to produce drawings at the same rate as on the drawing board. Within one month, the draughtsmen can be expected to be fully conversant with the CAD system, although experience continues to develop thereafter. In our experience the selection of the right type of person has a major influence on the productivity improvements. Full-time CAD operators are generally significantly more productive than draughtsmen who share their time with a drawing board. Technicians/detailers/draughtsmen are generally more productive on CAD than are 'part time' engineers.

MR JONES (Paper 5)
In reply to Mr Young, the decision to invest in CAD was made on the basis of a multi-functional system, with particular attention given to the application of RC and structural steelwork detailing.

The RC detailing (RCDS) package required additional staff training following the initial GDS A and B courses, and the overall period of training courses covered twelve days, followed by intensive in-house practice. Productivity steadily increased in accordance with proficiency. The present output indicates a fourfold increase in drawing production, and in the case of RC work much of this has accrued from the automatic bending schedule production.

In reply to Mr Cooper, the matters raised in support of a CAD system are very relevant and are fully supported by the Authors.

The system in Dorset was introduced on the basis of the following hardware: one CPU with a memory of 5 megabytes; three workstations; one plotter; one hard copier; and one digitiser.

The capacity of the CPU memory was increased recently to 9 megabytes with four further workstations, and future expansion will include a further CPU and two workstations, together with upgrading of the present disk drives from 75 megabytes to 170.

Minicomputers may not have the power required by CAD systems but, as they are used specifically as design tools in the Transportation and Engineering Department, we envisage that there will be a need for minicomputers in the foreseeable future.

In reply to Mr Nicholson, when introducing a CAD system, it is of utmost importance properly to identify and undertake the associated staff training requirements. With regard to RC detailing, each proposed user undertook the basic GDS A and B training courses, followed by two RCDS courses, making a total of twelve training days. This was followed by a period of in-house training and practice. Training costs per member of staff were in the region of £2000, and an average figure for proficiency in use of the system would be approximately three months.

MR ROWLEY (Paper 6)

In reply to Mr Young, it has been our experience that productivity gains are largest when producing details similar to those already stored in the CAD system. Exact copies are easily dealt with by manual draughting and, therefore, do not produce large productivity gains on CAD. Effective productivity gains can be obtained when producing similar details through CAD editing commands, but gains are largest if the details can be drawn automatically from parameterised information. It is, however, also our experience that details suitable for parameterisation are relatively few and far between in the civil engineering industry.

In reply to Mr Cooper, the suitability of minicomputers for CAD depends on several factors. One of these involves the capabilities of the graphics workstation and sophistication of the CAD software. For example, Tektronix 41XX and 42XX workstations are capable of accepting amendments to the stored picture in terms of content and proportion shown on the display. Not all software is capable of using these features and will use the mini-computer and communications resources to a greater extent. However, of greater significance is any statement made by the vendors as to the capacity of the mini-computer to handle a given number of workstations. To put fewer workstations on the system increases individual throughput, but at a higher cost per seat. In our opinion, minicomputers can be bought at sufficiently low cost to give satisfactory performance for 2D CAD. The resources required by RC detailing and 3D modelling, are far greater, however, and the workstation approach becomes more obviously worthwhile.

There is a definite trend towards a blurring of the boundaries between microcomputer and workstation, and workstation and minicomputer. We believe that, within the suggested timescale, each user will be equipped with dedicated computer power which would include his/her own main and graphic processors. On some sites the disk storage will be distributed to each workstation and opportunities to share data will be limited. It is to be hoped that most sites will be able to justify a fully networked system, with centrally managed documentation and archiving of graphic data.

In reply to Mr Nicholson, at WS Atkins the training of technicians in the use of CAD RC detailing has been carried out only after they have become fully proficient in general draughting by CAD. Having mastered the basic system, the learning of an additional module, such as RC detailing, is a relatively simple matter and we would estimate that two days' training together with two weeks' use would be adequate to achieve proficiency. Learning times can vary considerably between operators but, on average, we would expect training in general draughting to last two weeks and that operators would become proficient after three months' full-time use.

MR G. K. RAVEN, Ward Building Systems Ltd

As a fabricator, we have the BOCAD system mentioned by Dr Watson for detailing structural steelwork. Although we have arranged suitable hardware, with an optic fibre link into the works, there is, at present, no direct link between the CAD and the NC machines. This is currently being developed.

With regard to the question of 2D or 3D systems, our experience is that gains from pure draughting and scheduling from 2D work are not generally available. To obtain full benefits, 3D systems are essential to give an integrated method of working and to minimise checking and routine tasks at the terminal. Drawings are used as a means of communicating information, and the larger systems have the potential to provide this function without the need for checking each individual drawing. Only the basic input model needs to be checked both in-house and by consulting engineers once the system has been established. This will clearly reduce work for all parties and will speed up a difficult area in the present organisation of construction.

MR E. O'BRIEN, Electricity Supply Board
How close to reality is the concept of linking computer aided design and detailing with the manufacture/fabrication of structural steelwork?

The trend in reinforced concrete seems to be towards the production of detail reinforcement drawings and schedules, whereas in structural steelwork the trend is towards 3D modelling and subsequent analysis rather than detailing. Could the Authors please comment?

DR WATSON (Paper 7)
In reply to Mr Raven, I agree that there is limited advantage in using 2D systems for detailing structural steelwork.

Once drawings become no more than representations of a more comprehensive model held in the computer, their role as the primary means of communication is indeed threatened. However, technical and organisational problems hinder the transfer of such models between computer systems and will delay widespread communication by this means.

Integrated systems offer substantial potential for a reduction in the need for checking, but this may be better viewed as a potential to reduce the number of errors. It is essential that the industry is alert to the consequences (and dangers) of such technological changes. Contractural relationships may well have to be revised to compensate for these changes.

In reply to Mr O'Brien, much of the technology needed to link CAD and CAM for structural steelwork exists already, and specific localised examples can (almost) be found now. However, we are probably 15 years away from its widespread application to a broad spectrum of products. The difficulty is to achieve sufficient flexibility in the integration of systems and a cost that is economic. These issues are being addressed by the Eureka research project EU130, which seeks to establish the foundations for structural steelwork CIM systems. This pan European project started recently and is expected to last four years.

Comparisons between reinforced concrete and structural steelwork are interesting since the latter appears to have substantially greater potential for automation of the manufacturing process. It may be that we will see greater use of pre-cast concrete in the future. I agree with the trend described for reinforced concrete but suspect that the trend is more complex in the case of steelwork. Not all structures need an analysis based on a 3D model while detailing is frequently not performed by the designer. A major result of EU130 should be a "product modeller" for structural steelwork that will allow the required information to flow between design, detailing and manufacture.